OP

LIVING TOGETHER IN THE SEA

BY DR. LEON P. ZANN

Front Cover: *Nomeus gronovii* and *Physalia physalis.* Photo by
Charles Arneson.
Frontispiece: *Amphiprion perideraion* in sea anemone. Photo by
Dr. Leon P. Zann.

*Photographs and illustrations by the author unless
otherwise indicated.*

ISBN 0-87666-500-8

© Copyright 1980 by T.F.H. Publications, Inc. Ltd.

Distributed in the U.S. by T.F.H. Publications, Inc., 211 West
Sylvania Avenue, PO Box 427, Neptune, NJ 07753; in England by
T.F.H. (Gt. Britain) Ltd., 13 Nutley Lane, Reigate, Surrey; in Canada
to the book store and library trade by Beaverbooks Ltd., 150 Lesmill
Road, Don Mills, Ontario M38 2T5, Canada; in Canada to the pet
trade by Rolf C. Hagen Ltd., 3225 Sartelon Street, Montreal 382,
Quebec; in Southeast Asia by Y.W. Ong, 9 Lorong 36 Geylang,
Singapore 14; in Australia and the South Pacific by Pet Imports Pty.
Ltd., P.O. Box 149, Brookvale 2100, N.S.W. Australia; in South Africa
by Valid Agencies, P.O. Box 51901, Randburg 2125 South Africa.
Published by T.F.H. Publications, Inc., Ltd, the British Crown Col-
ony of Hong Kong

Contents

Introduction

Partnerships between different species of animals and plants hold a special fascination for men. It may be because they are uncommon biological curiosities and exceptions to the commonly held 'dog-eat-dog' concept of Nature. Perhaps it is because man can identify with the partners, especially the exploiters such as the ants which farm and milk aphids, the clams which farm and harvest algae, and the fish which are reputed to lead sharks to their prey.

Those species which closely cooperate with each other are particularly interesting. Birds and fishes remove and eat parasites of other animals; microorganisms digest the plant material consumed by termites and grazing mammals; and microscopic plants are biochemically and physiologically linked with other plants or animals. Man himself is involved in many mutually advantageous partnerships with other species. The first of these was with the dog, once a hunting companion but now mainly a psychological one. He later domesticated cattle and sheep and crop plants, taking over their evolution and molding them more closely to his needs.

The partnerships in which one member harms the other, parasitism, are regarded with a certain repugnance. Many plants and animals parasitize man, and in the underdeveloped countries of the Third World they are still major causes of disease and death. Yet parasites themselves are exploited by man to control harmful pests. In Australia, rabbit plagues have been controlled by spreading myxomatosis, a mosquito-borne parasite, and vast tracts of grazing land rendered useless by the spread of cactus have been regained following the introduction of an insect parasite of the plant. Small parasitic wasps are now commonly used to control various insect pests throughout the world.

The study of partnerships is an ancient one. The Greeks and Romans of antiquity, the first natural historians, described several examples. They found small pea-sized crabs in the Mediterranean oysters and mussels they ate, and they were intrigued by the pilot-fish and suckerfish which escorted sharks and their own lumbering sailing craft and galleys. Herodotus, the Greek historian, described an unlikely association between the Egyptian plover and the giant Nile crocodile. This bird habitually enters the gaping mouth of the reptile to feed on parasitic leeches on its gums.

The new natural historians of the nineteenth century began a special study of partnerships. In 1879 the French naturalist de Bary coined the word *symbiosis* (meaning 'living together') to describe the intimate and regular association between different species. The active partners were called *symbionts*, or sometimes *symbiotes*, and the study was called *symbiontology* or sometimes *symbiology*. The passive partner was called the *host*. The term *symbiosis* had a general meaning, with other words coined to describe the different types of partnerships.

Commensalism was coined to describe partnerships involving the sharing of food and *mutualism* was coined to describe mutually advantageous partnerships. Terms less frequently used include *endoecism*, which describes partnerships in which one animal shelters in the burrow of another; *inquilinism*, which describes those partnerships in which the symbiont shelters in the host; and *epizoism*, which describes those partnerships in which the symbiont lives on the host's surface.

Unfortunately the terminology is now in a state of confusion as different biologists use the same terms to describe different phenomena. *Symbiosis* has been used to describe mutually advantageous associations only, or sometimes those mutually advantageous associations at a physiological or biochemical level. *Commensalism* is often used to cover all associations that are not parasitic or mutually beneficial.

The breakdown in the terminology is not solely due to misunderstandings among biologists. It is also due to their failure to realize that biological processes are continuous and multifaceted and defy 'pigeonholing.' A symbiont often gains several advantages, for example shelter and protection, respiratory currents, food, etc., and the host might or might not be harmed on different occasions. Should the partnership be described as inquilinism, commensalism or parasitism? A third reason for confusion is that an association has to be very closely investigated before its nature can be established, but most associations have not been closely studied.

10

The terminology has therefore been used guardedly in this book. *Symbiosis* is used in its original and general sense and the other terms are discussed in Chapter One. The pedantic discussion of terminology does little to further this interesting and exciting subject which to me epitomizes the complexities, intricacies and mysteries of Nature and demonstrates the driving power of natural selection . . . and nowhere are there more or more varied partnerships than in the sea.

My own interest in symbiotic associations evolved very gradually. Like other students, I had been intrigued by the lessons on the quaint and unlikely partnerships between species, but the interest was the interest that we all have in freaks. It was many years before I realized that symbiosis was not merely a curiosity, a chance combination, but a very important biological phenomenon, especially in tropical waters.

The realization of its importance began during a study I conducted on marine fouling in tropical seas. Like all maritime institutions, the Royal Australian Navy had been troubled for many years by the failure of their antifouling paints to prevent the settlement of barnacles, tubeworms, bryozoans, oysters, algae, sponges and other sessile organisms on vessels and installations in the tropical seas. They therefore instigated my study of the biology of fouling.

One of my approaches to the problem was to investigate how marine reptiles and mammals antifouled themselves. I reasoned that as they had dead skins free of secretions—unlike that of the fish—they should be colonized by many sessile organisms. I therefore examined a range of animals and was surprised to find them free of all fouling organisms except for a number of aberrant barnacles. These barnacles tended to be host specific: one species was found only on the tails of certain species of sea snakes, another lived only on turtles' flippers or on dugongs' backs, another lived only on the shells of turtles, another lived only in their gullets, and others again were specific to the various whales and dolphins. I found myself wondering how the microscopic larvae managed to locate their various hosts in the great expanse of the sea. How did they distinguish a turtle from a sea snake? How did they settle on a whale as it surged along? What were the advantages of having such specialized ecological niches?

I found my interest being transferred from the hosts' antifouling mechanisms to the symbionts themselves. In my spare time I began to look for other examples on the fringing coral reefs near my home and on the patch reefs of the Great Barrier Reef which I regularly visited. I scanned a spectrum of the reef community—the corals, worms, molluscs, echinoderms, the fishes and others—spending several months on each group.

Little by little I began to recognize that, far from being a curiosity, the phenomenon of symbiosis is very common in the coral reef community and has a special importance. Perhaps a quarter or more of all coral reef species are symbionts. In fact, the coral reef is built on a partnership: the reef-forming or hermatypic corals are actually an intimate combination of plant and animal. In addition, a single colony of the coral might host as many as two or three dozen species of symbionts. It has been suggested that the thousand or so species of fishes living on a reef of the Great Barrier Reef might host as many as fifty thousand species of parasites, and most of these fishes rely on a behavioral symbiosis with cleaner-fishes for ectoparasite control.

Partnerships were also surprisingly common in certain open-water and soft-bottom communities I examined, and my background reading indicated that they are also found in many other communities in all of the world's seas. The ecological importance of symbiotic associations is probably yet to be recognized.

This book is in no way intended to be a definitive work on the subject. Rather, it is intended to be a stimulating introduction for students, aquarists, naturalists and biologists to a fascinating aspect of biology. A wide range of symbiotic associations are cited and discussed in detail wherever possible, but little is known about most of the examples cited. Extensive lists of species names are avoided to make the text more readable. Special emphasis is placed on the pictorial material, as written descriptions cannot adequately convey the intricate morphological adaptations and beauty of the associations.

The text is divided into five parts: (1) the biology of symbioses; (2) the occurrence of symbioses in different ecosystems; (3) the symbionts of certain hosts; (4) physiological mutualism; and (5) vertebrate symbioses.

Although one chapter is devoted to parasitism, the major portion of the book is concerned more with non-parasitic symbioses. Many books have been written on parasitism and the subject obviously cannot be adequately covered in a single chapter.

I have endeavored to examine at first hand as many of the associations as possible, and my observations and speculations are clearly indicated in the text. Other information comes from a variety of sources, the major ones being listed in the chapter-by-chapter bibliographies.

Acknowledgments

I am very grateful for all those who collected, photographed and identified material for me, especially my friend and colleague R.A. Birtles, and Walt Deas, one of the world's leading underwater photographers, who supplied some superb photographs.

Photographs were also supplied by Roger Steene, a leading Australian underwater photographer, and Dr. L. Cannon (parasites). Biological material was supplied by Dr. G. Allen, Dr. M. Borowitzka, Prof. W. Dunson, Dr. G. Heinsohn, R. Kenchington, C. Limpus, I. Loch, Dr. J. Veron, L. Zell and many others. Identifications were made by R. A. Birtles (echinoderms), A. H. Banner (pistol shrimps), W. Dowd (fish), Dr. J. Haig (half crabs), Dr. P. Gibbs (polychaetes), Dr. W. Ponder (molluscs), E. Pope (barnacles) and Drs. J. Veron and T. Done (corals).

Dr. M. Borowitzka contributed the chapter on symbiotic algae, his specialty.

Dr. T. Done, my sister Maria and others proof-read the manuscript and suggested numerous changes. M. Richards typed the manuscript.

Much of the research was conducted during free time while I was employed at the James Cook University of North Queensland, and I gratefully acknowledge all of those at the Department of Marine Biology who assisted me. Research was conducted on the R.V. *James Kirby* (James Cook University), Heron Island (Great Barrier Reef Committee), Lizard Island (Australian Museum), along the southern and middle Great Barrier Reef (Commonwealth Lighthouse Service: M.V. *Cape Moreton*) and the northern Great Barrier Reef (Royal Society and Universities of Qld. Great Barrier Reef Expedition).

Petrolisthes maculata, one of the porcelain crabs (which are intermediate between shrimps and true crabs), lives quite comfortably among the tentacles of sea anemones. Photo by G. Marcuse.

PART I. THE BIOLOGY OF SYMBIOSIS

Chapter 1. Non-Parasitic Associations

Symbiosis is a general term lumping together a whole conglomeration of partnerships which have evolved under different conditions involving different species. The details of these partnerships vary to such an extent that it could be said that there are almost as many categories of symbiosis as there are numbers of partnerships, for it is axiomatic that no pairs of organisms can be identical.

Like human partnerships, animal partnerships vary in the level at which they occur as well as in their fundamental details. Human partnerships operate at the level of international treaties through to business partnerships and the partnerships between man and woman. Animal partnerships range from ecological or behavioral ones through to physiological and inter- and intra-cellular ones. The extent of the accruing benefits and the degree of dependence is likewise variable. In certain partnerships both partners benefit and in some cases neither can survive without the other. More often only one partner benefits although the exploited partner may not be harmed. But in many cases, where the association is more intimate, one partner may be harmed and may even die because of the other.

Partnerships in which the hosts are harmed are termed *parasitism* and are the topic of the next chapter. This chapter outlines the other types of partnerships and examples discussed later in the text are briefly mentioned.

The difficulties in categorizing symbioses and the confusion in the terminology have been stressed in the *Introduction,* but there is still a basic need for some grouping, even if it is artificial and imprecise. The terminology adopted here is a common, but not a universally accepted, one.

INQUILINISM AND ENDOECISM: ASSOCIATIONS FOR PROTECTION

Animals in the sea have a wide range of mechanisms for defense and offense. Snails have thick shells and close-fitting doors or opercula to close themselves in. Sea urchins have arrays of sharp and sometimes venomous spines, and anemones and jellyfishes have batteries of stinging cells called nematoblasts.

Fishes such as barracudas and sharks have few enemies because of their size and teeth. Other fishes may rely on their speed or camouflage, while many small, vulnerable fish gather in schools for protection. There is some degree of safety in numbers, for predators may be confused by a milling or exploding school. Poisonous, toxic and unpalatable organisms are avoided by predators and often make themselves conspicuous so they are not accidentally taken.

The specialized defenses of many species are exploited by defenseless species seeking protection. Boldly striped pilotfish swim near oceanic sharks, formidable allies for any little fish. Juvenile pilotfish, trevallies (jacks) and some adult driftfish shelter near or even among the venomous tentacles of jellyfishes. Likewise, the little clownfishes shelter among the tentacles of venomous giant anemones and clingfishes live among the arms of unpalatable crinoids.

These partnerships are not physically close, but the defenseless animal obtains a degree of *protection by proximity*. Associations in which one species regularly shelters near, on or even within the body of another species for protection are termed *inquilinism* (from incolinus: 'who lives within').

Protective associations have evolved in many different ways under different ecological pressures. In many cases the ancestors of the symbiont and host could have been ecologically similar and the two might have lived in close proximity. For example, most of the cardinalfishes and long-spined diadem sea urchins tend to be nocturnal and shelter in caves and crevices during the day. Free-

16

Cardinalfishes (here *Cheilodipterus* sp.) often shelter among the spines of the sea urchin *Diadema*. (4m. Magnetic Island)

living cardinalfishes may often be seen swimming close to or even among the spines of the diademids, and it seems almost inevitable that certain cardinalfishes should form regular associations with them. Some of these sheltering cardinalfishes may also feed on host material such as mucus, feces and tubefeet to supplement their diets. Occasionally they also feed on their hosts' parasites and predators, thereby benefiting the urchin.

Certain of the refugees have an even closer physical association with their hosts, either living on them or within the body cavity. The transparent pearlfishes and messmates live in the mantle cavities of large tropical bivalves or in the cloacas and large coelomic cavities of sea cucumbers. Many shrimps, crabs and copepods also live in the cavities of bivalves and ascidians.

Like the cardinalfishes which feed on their protectors, many of the symbionts which live on or within their protectors also feed on them or steal their food. These symbionts may be literally standing on a food source, for example the host's mucus or the food the host has trapped for itself. Thus by natural selection over many

17

generations certain inquilines with unspecialized diets have proceeded along a path to commensalism.

The most specialized of the inquilines are those which modify the growth of their hosts to create a nest or gall for themselves. Certain small copepods and crabs alter the growth of their sea urchin hosts so that they become encased in a protective gall. Likewise, the gall crabs and gall shrimps modify the growing tips of certain branching corals so they form a cell which ultimately imprisons them within.

Dromid (sponge) crabs and some majid (spider) crabs shelter under sponges and other invertebrates which live in the same areas. Some sponge crabs carry whole sponges but others carve them into a neat-fitting cap which they carry on their carapaces. The camouflaged majid crabs nip off pieces of the surrounding organisms and 'plant' them on their backs where they are held in place by long hairs. These crabs discriminate in their selections for camouflage and use the different organisms in much the same proportions as they are present in the surrounding community. If the crabs are transferred to a different community they will discard their old camouflage and plant a new one on their backs. Some camouflage crabs regularly plant sea anemones on their carapaces; the anemones have the added advantage of being venomous, thus deterring predators. Likewise, many of the hermit crabs plant sea anemones on the snail shells which they wear to protect their soft abdomens.

ENDOECISM: SHELTER IN THE BURROW OF ANOTHER

A variety of bottom-living or benthic invertebrates from most major phyla dig burrows or produce tubes so they may be protected from predators and the environment. Some of these excavate tunnels in mud, sand or even rock, while others secrete a calcareous or organic tube for protection. Burrows range from small temporary ones to large permanent ones with many chambers. The latter are particularly attractive to animals seeking shelter. The term *endoecism* is applied to those associations in which the symbiont, the endoekete, habitually shelters in the burrow or tube made by another species.

18

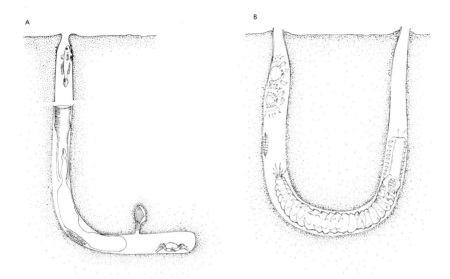

The burrows and tubes made by worms give shelter to a number of lodgers. (A) The burrow of the echiuroid worm *Urechis* is inhabited by arrow gobies *(Clevelandia),* pea crabs *(Scleroplax),* a scaleworm *(Hesperonoe),* and a bivalve *(Cryptomya).* (B) Likewise, the tube of the polychaete worm *Chaetopterus* is occupied by porcelain crabs *(Polyonyx),* and sometimes a polynoid scaleworm. (Redrawn from MacGinitie and MacGinitie, and Gotto.)

On many occasions I have walked along sandy or muddy beaches and startled the gray ghost crabs or festive fiddler crabs. In the scatter for shelter some invariably dart into the nearest burrow. They may be promptly evicted by the resident crab or may remain there until danger passes. Cases of endoecism could have arisen in this way and become more refined as time progressed.

In North America the arrow goby, *Clevelandia ios,* habitually occupies the burrows of various invertebrates of the sandy and muddy estuaries. At the approach of danger or as the tide ebbs they seek refuge in any hole available. The burrows of echiuroid worms are preferred, and a single burrow may house as many as two dozen fish as well as a pair of pea crabs.

The enterprising gobies have an interesting relationship with their fellow lodgers, for if they find particles of food too large to swallow they will give them to the crabs. The crabs proceed to eat them, shredding them in the process, and the gobies then take back some of the smaller fragments.

Filter-feeding bivalves also live in the shelter of the echiuroid burrows and benefit by the feeding and respiratory currents set up by their hosts. In addition, small polychaete scaleworms also share the burrows and feed on their hosts' wastes.

PHORESIS: ASSOCIATIONS FOR TRANSPORT

Marine animals may be classified according to their mobility. Some drift passively in the water column (plankton) and others are active swimmers (nekton), while others float or sail on the surface (pleuston). On the sea floor some benthic organisms are sessile, permanently attached during their larval or, more often, during their adult lives. Others are mobile and may either walk with legs, crawl or wriggle.

As a result of processes of evolution some bottom-dwellers such as pteropod snails have taken off into the waters above, while some free-swimming animals have lost their mobility to become sessile, for example the garden eels, which are relatively immobile because of their dependence on burrows.

Such changes in mobility entailed considerable morphological alteration, but many sessile groups seem to be too rigid and specialized to change. For instance, there are no swimming sponges or barnacles, although both have mobile larvae.

However, sessile organisms may become mobile by exploiting mobile ones, and various essentially benthic organisms are then able to exploit the waters above. The barnacles, for example, may seem rigidly modified for a benthic way of life, but they have ventured into the water column on the backs of mammals, reptiles and even jellyfishes. Many also live on tree-like sponges and corals, thus exploiting the waters immediately above the bottom.

Associations for transport have been termed *phoresis* (from *pherein*: 'to carry'). This transport may enable the recipient to feed in new areas, to exploit the feeding and respiratory currents generated by the host's movement, to move with little expenditure of energy, to avoid predators, to capture prey and to disperse their young.

The barnacles are prime examples, but the pilotfish and suckerfish also use their hosts for energy-conserving transport as well as for food and protection.

EPIZOISM: ASSOCIATIONS FOR A SURFACE

The sea can be very crowded. A visitor to a rocky coast cannot fail to notice the densely populated bands of oysters, tubeworms, mussels or barnacles. Alternatively, the sea, like the land, can be featureless and sparsely populated. This is most apparent in the vastness of the open sea or in the expanses of mud flats of estuaries. In over-crowded communities organisms may grow on others simply because there is no other available unoccupied space. By contrast, in the featureless substrate-limited environments organisms may settle on others because they offer a surface for settlement where few are otherwise available. In the crowded coral reef communities barnacles may settle on living corals. On muddy shifting bottoms barnacles may settle on sponges and hydroids and on mobile crustaceans such as crabs and lobsters. In the open sea they may settle on jellyfishes or even whales.

Associations in which animals live on others for substrate requirements are termed *epizoism*, and the symbionts are called

A dense band of mussels on a rocky New Jersey shore. The competition for space has led some animals to seek out others as a substrate. Photo by Dr. Herbert R. Axelrod.

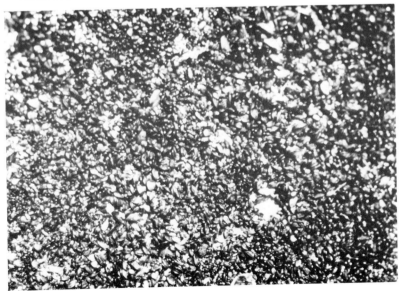

epizoites. In many cases the symbiont also gains food, currents, and transport as well and there is an overlap with phoresis and commensalism.

COMMENSALISM: ASSOCIATIONS FOR FOOD

There is an element of food stealing in many of the above symbioses. The crabs in the worm tubes may take mucus from the host, the suckerfish and pilotfish may take scraps from their hosts' meals and the shrimps and pea crabs sheltering in the mantle cavities of large bivalves may steal the plankton the hosts have trapped in their gills.

Associations in which food is shared or taken by one partner from the other without greatly harming it are termed *commensalism* (meaning: 'at table together'). This term is also used to describe all non-parasitic and non-mutualistic associations and may thus simplify the difficulties in terminology.

The large well-protected mantle cavities of bivalves such as this *Pinna* (opened) offer shelter to many refugees. Commensals steal food from the exposed gills. (Intertidal, Magnetic Island)

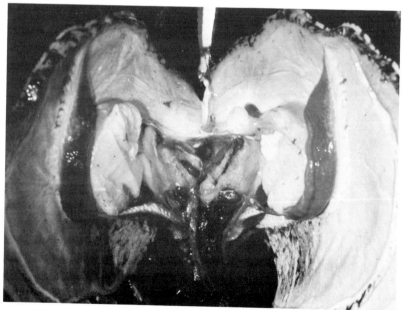

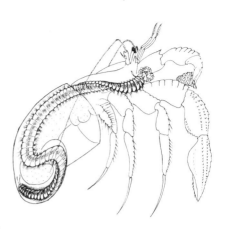

A commensal polychaete worm *Nereis fucata* steals food from the meal of its host, the hermit crab *Eupagurus bernhardus*. The crab has been given a glass shell, enabling the worm to be seen. (Redrawn from Caullery, after Thorson)

There are a few cases of one partner voluntarily giving food to the other. Anemonefishes, particularly those in aquaria, may take food and deliver it to the tentacles of their host, and some hermit crabs may take a piece of their food and place it in the tentacles of their anemone. A coral crab has the opposite relationship with the small anemones it carries in its claws: it uses these anemones to capture food and then takes some of it for itself.

Some commensals steal their hosts' food. One daring polychaete worm that lives in the shell occupied by an Irish Sea hermit crab actually snatches food from the jaws of the crab. Others may eat their hosts' metabolic wastes and feces. A female crab that lives in the rectum of sea urchins eats undigested food and intestinal flora in the feces. Many shrimps, crabs and copepods live on the surface of corals and other coelenterates and eat mucus, adhering detritus, dead skin and the expelled symbiotic algae, the zooxanthellae.

A large and diverse population of commensals may live on a single host and take shelter, protection and food from it without doing damage. Certain organisms are particularly prone to exploitation by commensals because of their feeding habits and physiology. Predators which bite chunks from their prey and scavengers that shred their food externally lose fragments into the surrounding water, and these may be taken by waiting commensals. Commensals also pirate food from the plankton feeders which either pump water through sieves, spread or sweep nets through the water or trap food on mucous nets. The soft-bodied

23

coelenterates and molluscs which secrete copious amounts of mucus to keep their bodies moist and free of silt are also prone to exploitation by mucus-eating commensals.

It has been stated that during the course of evolution partnerships which initially evolved for protection, shelter, transport, feeding currents or a surface for settlement might have shifted to primarily commensal ones. A symbiont which hides on a mucus-secreting host or one that is a messy feeder may begin to take food from it.

If a source of food is freely available it seems almost inevitable that some animal will exploit it. Many animals are surprisingly opportunistic and plastic in their feeding habits (captives in zoos and aquaria often thrive on substitute diets). Thus an ancestral symbiont hiding on its host might have supplemented its normal food with the host's own food or its wastes.

In certain cases the differences between commensalism and parasitism are diffuse. A commensal feeding on dead cells may inadvertently take some living material; commensals living on the host's body surface may irritate it, resulting in increased mucus production and calloused tissue. If the burden of commensals is too high the host may expend too much material and energy in the tissue repair and production of mucus and its health may suffer.

MUTUALISM: RECIPROCALLY ADVANTAGEOUS PARTNERSHIPS

In most of those partnerships described above one animal (the symbiont) gains something, while the other (the host) gains nothing. But in certain cases the host may also gain from the association, sometimes to the extent that neither partner can survive without the other. Such mutually or reciprocally advantageous partnerships have been termed *mutualism*. In the past *symbiosis* has been used as a synonym, especially in cases where there is a physiological exchange between partners.

Mutualism is often considered to be the most interesting of all symbioses for it can entail very basic and intimate interactions between the most unlikely of organisms. The special attributes of each partner may be combined so that the pair becomes a kind of 'super-organism'.

24

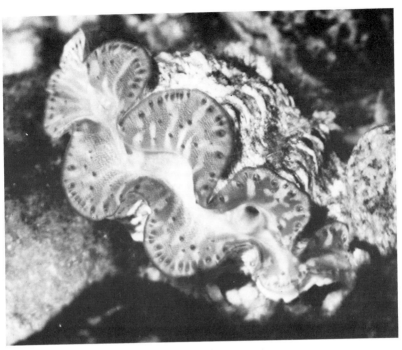

The giant *Tridacna* has formed a close physiological alliance with single-celled algae called zooxanthellae which it 'farms' in its exposed mantles. Photo by G. Marcuse.

Thus the hermatypic corals which build vast structures in tropical seas have escaped from the boundaries of ordinary coelenterates by forming physiological alliances with single-celled algae called zooxanthellae. So too have the giant *Tridacna* clams which farm pastures of zooxanthellae in nutrient-bathed 'greenhouses' in their mantles. Certain deep-water cephalopods and fishes similarly culture photogenic bacteria in light organs which they use for illumination when hunting or for the avoidance of predators. Mutualistic microorganisms living in the guts of many herbivorous invertebrates and vertebrates assist in the breakdown of their hosts' food, enabling them to exploit food resources that they are otherwise not physiologically able to digest.

Mutualistic symbioses range from these intimate physiological partnerships to ecological and behavioral ones. A small solitary coral without natural powers of locomotion is able to live on soft muddy bottoms because of an alliance with a sipunculid worm that drags it around and keeps it from being smothered by sediment. The coral gains mobility and the worm gains protection. Similarly, a nearly blind pistol shrimp gains a pair of keen eyes from the alert goby that shares its burrow.

Cleaning symbiosis is a very basic and mutually beneficial interaction in many fish communities, particularly those in the tropical seas. Certain small specialized fishes and shrimps remove parasites, food wastes and diseased and injured tissue from other fishes, thereby gaining food for themselves and benefitting the health of their hosts. The acquisition of complex innate behavioral characteristics by the host fishes suggests that the symbiosis has had an important role in the success of the species.

Paradoxically, mutualism is likened to parasitism in that it may involve very basic biochemical, physiological and cellular interactions between different species. Mutualism can even be envisaged as a type of two-way parasitism in which each partner is a parasite of the other, taking from it and living at its expense.

Chapter 2. Parasitism

Almost every vertebrate, terrestrial or marine, is the host of a complement of parasites. So too are many of the invertebrates, although in most cases their parasites are not well known. Indeed, parasitism is so common that it is rare to find any organism which is not parasitized or even to find classes of invertebrates with no members which have adopted a parasitic way of life.

WHAT IS A PARASITE?

The term parasite is a very familiar one, but it is impossible to produce an all-encompassing definition of it. The difficulties in defining it again stem from our desire to categorize processes which are continuous and multifaceted and which have evolved independently in many different groups on many different occasions.

A commonly accepted definition states that: *a parasite is an organism which lives in, on or frequently visits another, usually larger, organism (the host), nourishing itself at the expense of this organism but without destroying it as a predator destroys its prey.*

This and other definitions refer to the rather arbitrary criteria of harm to the host and the time spent on it. However, sometimes an unusually large population of commensals may damage a host and, similarly, ordinarily harmless commensals may damage their host when little food is available or when it is sick or injured. The temporal aspect is also inexact. Mosquitoes and blood-sucking flies are often called *micropredators* or *parasitoids* rather than parasites because they are still free-living and spend only a small part of their time on their host or prey. On the other hand, the blood-sucking ticks which spend a greater part of their lives on their hosts are generally regarded as ectoparasites.

Other definitions of parasites may be more precise. For example, some parasitologists consider that a metabolic dependence of the symbiont on its host is the only good criterion of parasitism. Other parasitologists regard an immunological reaction by the host as a better criterion of a parasitic association.

The parasites, even though they are a rather artificial and ill-defined group of organisms, do show many basic similarities. These are the result of convergent evolution due to the similarities of their ecological niches. They generally involve the simplification of those organs relating the animal to its environment and specializations of their digestive and reproductive systems made necessary by their new way of life.

SPECIALIZATIONS OF PARASITES

The parasites are among the most degenerate and highly specialized of all organisms, so much so that some of them consist of only digestive and reproductive tissue embedded within the bodies of their hosts.

Because they live either on or in another organism, their environments are much more stable than those of the non-parasites. As a result of this stability their sensory, locomotory and skeletal organs become redundant and progressively degenerate. On the other hand, there is a hypertrophy or increase in their digestive systems and of associated organs of attachment and feeding. The reproductive system also hypertrophies, for the ovaries produce prodigious numbers of eggs to offset the enormous wastage of young because of their difficulties in locating a host.

The convergent evolution of many of the parasites is well illustrated by the blood suckers which include such diverse animals as flukes, leeches, a polychaete worm (*Ichthyotomus*), ticks, flies, copepods and others. In all of these there is an enlargement of the mouthparts for biting and sucking, the production of an anticoagulant to keep the blood flowing, an enlargement of the midgut for storage of blood and a reduction of the hindgut since the parasites' food contains few wastes.

The similarities are even more marked in the more degenerate of the endoparasites, which includes certain of the barnacles, the isopods, the copepods and the snails. These are remarkably similar for they consist of little more than a gonad and a membrane for nutrient absorption.

The adaptations among the parasities are manifold and generalizations about them can only be vague. It is therefore necessary to examine a range of the better known marine parasities and their particular adaptations to a parasitic way of life. In certain cases it is possible to trace the progressive evolution of these adaptations through a range of related but less specialized organisms, some half-way to parasitism.

PARASITIC WORMS

A number of endoparasites, belonging to several different phyla, have a simple worm-like appearance. These include the trematode and cestode flatworms (phylum Platyhelminthes), the round worms (phylum Nematoda) and the spiny-headed worms (phylum Acanthocephala). In addition to these, some of the parasitic snails are also worm-like.

Helminths

The phylum Platyhelminthes is comprised of three classes: Turbellaria, Trematoda and Cestoda. The first is mainly free-living but includes a number of commensals of echinoderms, crustaceans, molluscs and sipunculid worms. The others are exclusively parasitic and are so highly specialized that it is not possible to trace their evolution from their free-living ancestors.

Trematodes: Digenians

The digenians, or flukes, are trematode worms with flat bodies and oral and ventral suckers. The adults live in the body cavities—the gut, bile and pancreatic ducts, bladder, lungs and cranial sinuses—of vertebrates. Their bodies are flattened to lessen the resistance of passing host products, and they attach to their hosts with their oral and ventral suckers. Like their free-living relatives, the turbellarians, they have a simple body plan. Their gut is also relatively simple, but there is now evidence to suggest that they absorb food through their body walls as well. The adults

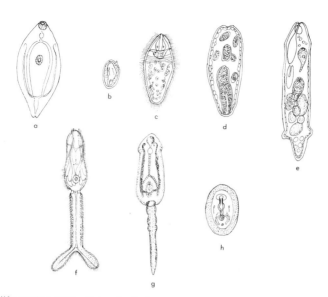

The life cycles of the flukes include a number of larval stages and hosts. (a) Adult fluke. (b) Egg shed into water. (c) Miracidium larva hatches and enters a gastropod. (d) Miracidium forms a sporocyst in which redia larvae develop. (e) Cercaria larvae develop inside redia. (f) Cercaria leaves the gastropod. (g) Cercaria enters a second intermediate host, or vegetates. (h) The vegetating metacercaria is swallowed by the definitive host. There it develops into the adult fluke.

have no sensory organs and their nervous systems are rudimentary.

The major adaptation to their parasitic way of life is in their life cycles. The adults produce a large number of eggs, but their reproductive potentials are further increased in a unique process known as larval multiplication or polyembryony—the larvae actually asexually produce further generations of larvae. This is extremely effective, and it has been estimated that a fluke may produce between 10,000 and 100,000 times more larvae than a free-living turbellarian.

Their life cycles are complex. Eggs are voided into the sea via the host's natural passages. In certain species these eggs may then fall to the bottom but remain undeveloped until they are accidentally ingested by a certain snail, the first intermediate host. The eggs of other flukes may hatch and the miracidium larvae actually seek out and burrow into their particular snail host and enter its digestive gland.

30

Life cycle of the seagull fluke *Stictodora lari*. (a) Egg leaves via the host's feces. (b) Miracidium larva burrows into the first intermediate host, the snail (c) *Velacumantus*. Many cercariae are ultimately produced from each miracidium. (d) The cercariae leave and burrow into a small fish such as a mullet. (e) They encyst in the muscles of the second intermediate host until such time it is eaten by a gull (f), its definitive host. The cercariae may burrow into man, causing lesions known as 'swimmer's itch'.

Life cycle of the fish fluke *Fellodistomum*. (a) Adult in gut of host. Eggs are shed into the water and hatch into miracidium which enter a first intermediate host, a bivalve (b). The miracidium develops into the redia larva (c) which produces many cercaria (d). The cercaria leave the redia and the bivalve and enter a second intermediate host, an ophiuroid (e). Inside the ophiuroid the larva develops into a metacercaria (f) which will develop into an adult fluke if its host is eaten by the definitive host, a fish (g). (Redrawn from Dogiel *et al.*)

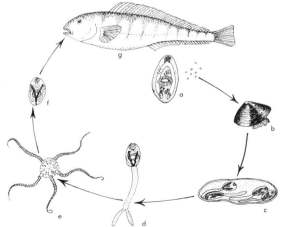

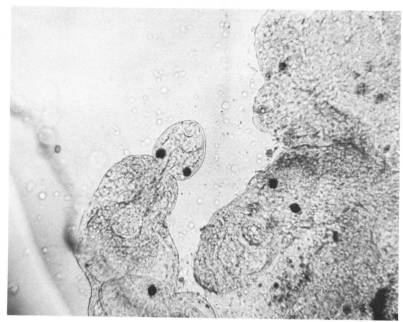

A redia larva of a trematode giving birth to a number of cercaria. This process of larval multiplication greatly increases the reproductive capacity of the flukes. Photo by L. Cannon.

In the host snail a larva changes into a second stage, the sporocyst, which produces many new larvae called rediae or daughter sporocysts. These produce yet another larval stage, the cercaria, which resembles tailed adults.

Cerceriae leave the snails and then either encyst on algae, actively seek out a second intermediate host in which to encyst or directly enter their final or definitive host. If the larvae encyst on a particular alga or if they encyst in a second intermediate host there is a chance that the alga or second intermediate host will be eaten by the final host.

The chances of the original larva finding its specific host are obviously very remote, tens of thousands or even millions to one. However, larval multiplication greatly increases the chances. A periwinkle kept isolated in a laboratory released an astounding 5,500,000 trematode larvae during five years. And after seven years an average of 1,600 larvae were still leaving the host every

The long, poisonous spines of the diademid urchins provide shelter for many fishes. Pempherids, apogonids and pomacentrids congregate in the protection of the spines of a long-spined sea urchin. Photo by Pierre Laboute. (New Caledonia)

day! Infestation of snails by trematode larvae may be very high in some areas and may even approach 100 %. In addition, a single snail may host as many as two dozen species of trematodes.

Cestodes: Tapeworms

The long, flat tapeworms which are intestinal parasites of all groups of vertebrates, with the exception of the crocodiles, are more specialized endoparasites than the flukes and show a simplification of the helminth body plan.

Tapeworms consist of an anterior segment known as the scolex which carries hooks and suckers for attachment and an elongate body or strobila composed of identical segments called proglottids. Each proglottid consists of an envelope through which the host's digested food is absorbed and reproductive tissue. All locomotory, sensory and nervous tissues have been lost.

There are many marine tapeworms, but little is known of their biologies or life cycles. It is known that the definitive hosts of the fish tapeworm *Bothriocephalus* include the turbot and certain other piscivorous fish. In this tapeworm a ripe proglottid packed with eggs is shed into the host's gut and passes out with its feces. In the sea the eggs hatch into ciliated coracidium larvae. If luck has it, these may be eaten by a certain small planktonic copepod. Inside the copepod's gut they shed their protective skin and burrow into its body cavity using special hooks. Once there they cease development and become dormant for as long as the copepod lives.

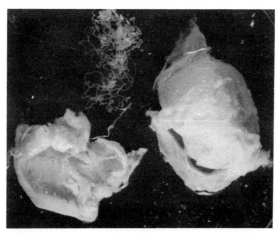

Cyst of the tapeworm (*Phyllobothrium chamissonii* removed from the flesh of the melon-headed dolphin *Peponocephala electra*. The adult tapeworm lives in sharks. Photo by L. Cannon.

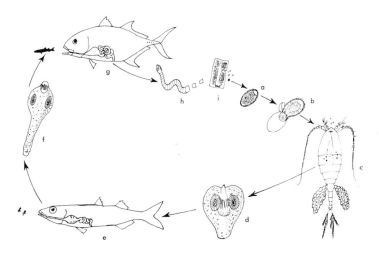

Life cycle of the marine tapeworm *Bothriocephalus.* (a) Egg from gravid pro-glottid shed into the sea in its host's feces. (b) Egg hatches into coracidium larva. (c) Coracidium is eaten by a copepod, its first intermediate host. (d) In-side the copepod the larva develops into a pleurocercoid larva. (e) If the copepod is eaten by a small fish the pleurocercoid attaches to its stomach wall and ceases development (f) until the second intermediate host is eaten by certain carnivorous fish (g). The pleurocercoid attaches to the definitive host's gut wall and begins producing proglottids. (From various sources).

However, if a small fish such as a herring eats the copepod the larvae are activated and attach to the gut wall of the fish. The young tapeworms then develop the adult scolex and a few proglot-tids but once again become dormant.

Only when the small fish is eaten by a larger one is the tapeworm reactivated. It then attaches to the gut wall of the sec-ond fish and completes its development.

The life cycle of this tapeworm seems even more precarious than that of the flukes. However, the adult tapeworms produce enormous numbers of eggs and the reproductive potential of each egg may be augmented by budding of the scolex of the larva to produce many new individuals. *Diphyllobothrium,* a tapeworm with a terrestrial host, produces as many as 7,000,000,000 eggs during its lifetime, and the marine species may be similarly fe-cund.

The helminth parasites are all hermaphoditic, though not self-fertilizing. This doubles the chance of a successful fertilization, for any two individuals may cross fertilize.

Slender suckerfish *Echeneis* and a golden trevally *Gnathanodon speciosus* sheltering near a giant Queensland grouper *Promicrops.* (20m. Bowl Reef)

The pearlfish *Onuxodon margaritiferae* shelters inside the mantle cavity of a pearl oyster during the day and feeds outside at night. (17m Lizard Island)

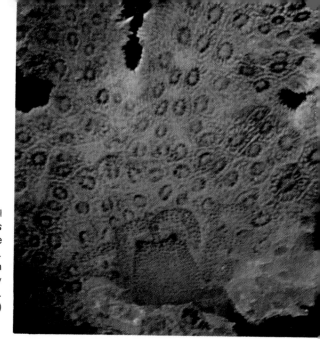

Juvenile female gall crab *Hapalocarcinus marsupialis* in its home in the coral *Seriatopora*. One wall of the prison has been broken away to show the crab. (Lodestone Reef)

The commensal shrimp *Anchistus custos* acquires protection, but also steals plankton and mucus from the gills of its bivalve host *Pinna*. (Intertidal, Magnetic Island)

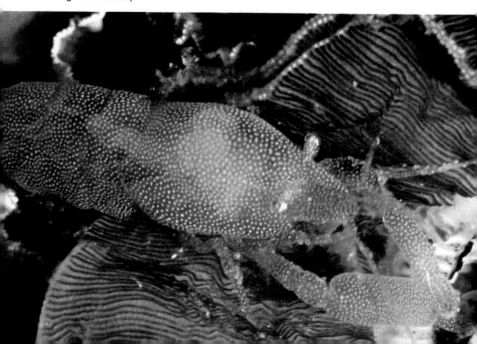

Nematodes: Roundworms

The roundworms are abundant in all ecosystems and include many free-living as well as parasitic species. The parasitic roundworms differ little from their free-living counterparts—they are round-bodied, elongate, have a terminal mouth and anus, external cuticle and longitudinal bunches of muscles. However, they are less mobile and lack the sensory apparatus and associated nervous systems of the others.

Unlike the helminth worms already discussed, it is possible to trace the evolution of parasitism in the roundworms; it is probable that they made this step to parasitism on at least four different occasions.

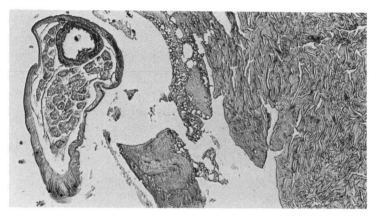

Section through the ear of the melon-headed dolphin *Peponocephala electra* showing a parasitic nematode *Stenurus globicephalae*. Heavy infestation of such parasites are thought to disrupt the sonar of whales leading to beachings or 'mass suicides'. Photo by L. Cannon.

Certain characteristics of the free-living roundworms make them well suited for intestinal parasitism. The adults have a tough cuticle resistant to their host's enzymes, an elongate shape and a high reproductive potential. In addition, the nematodes already have a complicated life cycle geared for dispersal of the young. This consists of five larval stages; the third, the dispersal stage, is extremely resistant to desiccation. The nematodes are an adaptable group and have exploited diverse habitats on the land and in the sea. It seems almost inevitable that they should also have become endoparasites.

PARASITIC CRUSTACEANS

The crustaceans are primarily a free-living group, and cases of parasitism are comparatively rare. However, in the groups Isopoda, Copepoda and Cirripedia parasitism has evolved on numerous different occasions and whole families may be parasitic.

The crustaceans and the gastropods are particularly interesting for they provide us with an insight into the evolution of parasitism and the progressive morphological degeneration this entails. This degeneration is most apparent in the crustaceans, which ordinarily show a complex and almost geometrical type of body plan. This is progressively simplified under the influence of parasitism until certain of the endoparasites are literally shapeless blobs. But for their typically crustacean larval life histories, these most degenerate parasites would not be recognizable as crustaceans.

Isopods

Isopods are small crawling crustaceans with a shell of plates rather like that of an armadillo. Most species are free-living, but some have become parasites of fishes and decapod crustaceans.

The gnathiid isopods, ectoparasites of fishes, do not show a radical deviation from the typical crustacean body plan, for only their larvae are parasitic. These attach to a fish and, after piercing its skin with their long mandibles, suck its blood. After several months on the host they metamorphose into adults and become free-living on the sea floor.

The cymothoid isopods are somewhat more specialized. Their larvae are free-living, but the adults attach to fish and suck their body fluids. One, the misshapen *Ichthyoxenus*, bores into the body walls of fishes and excavates a cavity inside.

Epicarid isopods are much more specialized and show gross alterations of the basic crustacean body plan. The larvae of the epicarids are free-living in the plankton until they encounter certain copepods to which they attach and feed on for some months. They later metamorphose into adults, leave their copepod host and seek out their definitive host, a decapod crustacean such as a shrimp or crab.The first epicarid to enter the host's brood pouch or gill cavity becomes a female, while latecomers become dwarf males which live parasitically on the female. (Dwarf and com-

A spider crab with sponges and algae 'planted' on the special hairs or setae of its legs and carapace. (35m. off Magnetic Island)

Long hooked hairs on the legs of majid carrier crabs are used for attaching camouflage. (30m. off Magnetic Island)

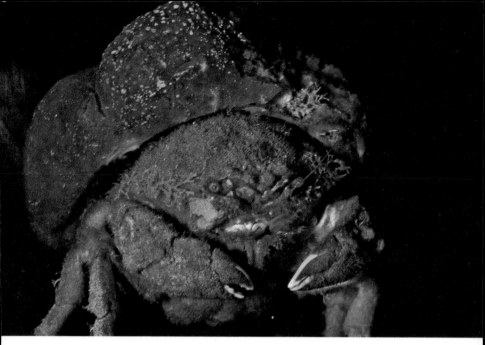

Sponge crab *Dromidiopsis edwardsi* holding a cap of sponge on its carapace. Note the encrusting bryozoans and algae adhering to the crab and damaged sponge. (35m. off Magnetic Island)

The rear legs of the dromid crabs are specialized for carrying sponges. (30m. off Magnetic Island)

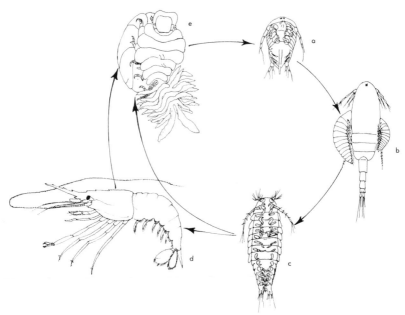

Life cycle of an epicarid isopod. (a) The egg hatches into a microniscus larva which attaches to a certain copepod (b). The microniscus develops into a cryptoniscus larva (c) which leaves the copepod and settles on its definitive host, in this case a shrimp (d). Inside the host the female larva matures into the adult (e). The male larva settles directly on the mature female and degenerates into testes. (Adapted from various sources).

plemental males are not strictly parasitic, for symbiosis is a relationship between *different* species. These cases are sometimes described as *intraspecific grafts.)* The males are very much reduced and are virtually only testes tissue attached to the female. The female herself has lost her sense organs, her legs are reduced so she can no longer crawl and her ovaries and brood pouch have increased greatly in size.

The most degenerate of the parasitic isopods are the cryptoniscids. These attach to their particular hosts and mature, firstly into adult males and then changing into females. These females gorge themselves on their host's blood and accumulate food reserves. Their digestive systems and other organs then atrophy until they are virtually only a large sac packed with eggs. When the sac bursts the eggs are liberated and the female—an empty sac—dies.

Copepods

Copepods are minute, usually planktonic, crustaceans. Parasitism is relatively common in the copepods, their hosts including almost all marine invertebrates and vertebrates—whales, fishes, ascidians, echinoderms, molluscs, worms, corals and other coelenterates. The parasitic copepods also demonstrate a progressive simplification of their body plan, from slightly modified blood-sucking ectoparasites to extremely degenerate endoparasites.

Copepods have a more complicated larval life cycle than other crustaceans: their larvae develop through six planktonic nauplius stages and then five copepodite stages. After this the free-living species begin their adult lives and the parasitic species seek their respective hosts. Despite radical changes in the morphologies of the parasitic adults, the larval life histories are little changed. The parasitic copepods range from the monstrillids, which parasitize polychaete tubeworms in their early stages, to the lernaeids, which progressively degenerate during their adult lives on two hosts.

This copepod lives in the egg sacs of the sand crab *Portunus pelagicus* and sucks out the contents of the eggs. Note the hole bored in the egg membrane. Photo by L. Cannon.

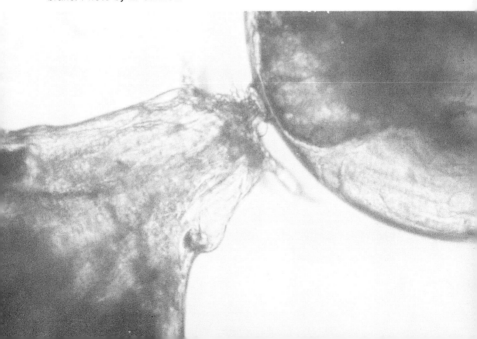

Burrows and tubes such as these made by a polychaete worm *Australonereis* may provide shelter for refugees. (Intertidal, Magnetic Island)

Soldier crabs *Mictyris longicarpus* form large aggregations on the tidal flats but disappear into sand burrows at the approach of the high tide. Photo by Keith Gillett.

The tubes of the polychaete *Chaetopterus* (here opened) provide shelter for other animals as well. (10m. Magnetic Island)

Polynoid polychaete worm *in situ* in opened *Chaetopterus* tube.

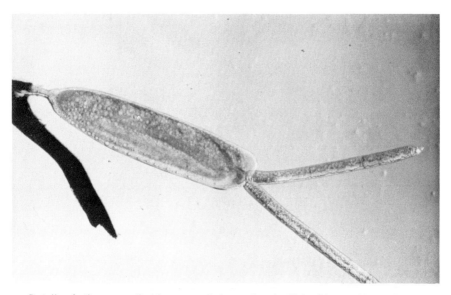

Detail of the cymothoid copepod from the batfish *Platax*. Note the characteristic egg sacs and the insertion of the head into a blood vessel near a fin ray.

Parasitic copepods. (A) *Pennella* on a flyingfish. Note the epizoic goose barnacles *Conchoderma* on the copepods. (Redrawn from Barnes) (B) *Lernaeenicus* buried in the retinal artery of its host. (Redrawn from Baer)

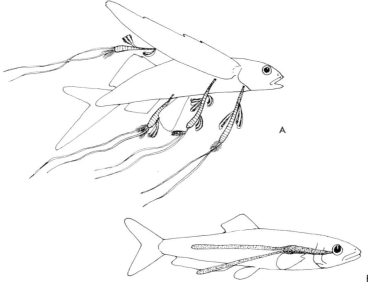

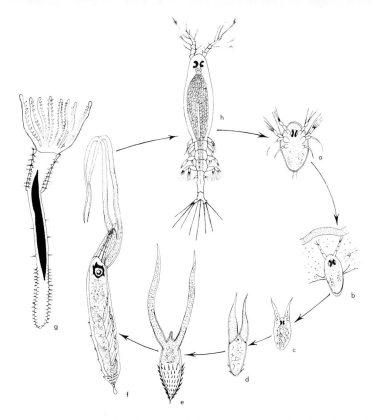

Life cycle of a monstrillid copepod *Cymbasoma.* (a) Nauplius larva. (b) Nauplius burrows into the body wall of a tubeworm. (c, d, e, and f) Stages in the development of the metanauplius and copepodite showing the growth of the two organs used for absorbing nutrients. (g) Position of copepodite larva in host. (h) Free-living adult lacks mouthparts and gut. (Redrawn from Baer)

Copepods of the families Lernaeoceridae, Pennellidae and Sphyriidae cause considerable damage to fishes. Adult females lose all appendages after they attach to the body surface of their hosts, grow through their body walls and attach to blood vessels or to their hearts. One of these, *Pennella*, parasitizes large pelagic swordfish and may reach 20 cm in external length. Only their abdomen and trailing egg sacs are visible, for the thorax and head lie permanently buried in the host.

The lernaeids have two hosts. Adult males and females initially parasitize a certain fish, but after fertilization the female leaves and settles on a second species. There she metamorphoses into a shapeless mass identifiable as a copepod only by the characteristic trailing egg sacs.

47

Some free-living animals have given up their mobility to gain protection from a burrow. These garden eels *Taenioconger hassi* have become relatively immobile because of their dependence on these burrows. Photo by Bruce Carlson.

Barnacles have formed associations with marine reptiles and mammals to gain transport. *Chelonibia testudinaria* is seen here on the green turtle *Chelonia mydas*. Photo by R.A. Birtles. (Raine Island)

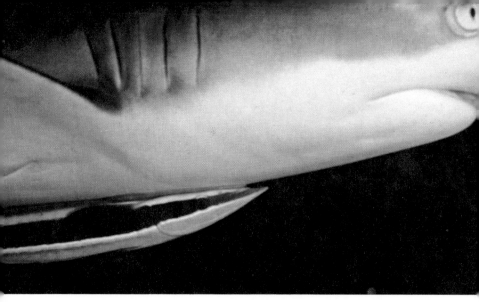

Suckerfish, such as this slender suckerfish *Echeneis naucrates,* hitch rides on large fishes—in this case a black-tipped reef shark *Eulamia spallanzini.* (Magnetic Island)

Trumpetfish *Aulostomus chinensis* may use other fishes for transport and cover when they stalk their prey. Photo by Walt Deas.

Cirripedes: Barnacles, Ascothoracids and Rhizocephalans

Barnacles, whose solid white shells are familiar to all visitors to the shore, are close relatives of the copepods and share with them five nauplius larval stages. Instead of metamorphosing into copepodite larvae, they develop into bivalved cypris larvae which seek surfaces for the adult stages to live on. When a suitable substrate is located they attach to it with a cement gland and then metamorphose into the adult. Unlike other crustaceans, adult barnacles are permanently attached by their backs to the surface, but they kick their legs (cirri) through the water to net planktonic food. They are encased in a shell of calcareous plates, a feature which led them to be classified with the molluscs for many years.

Parasitism has evolved on several occasions in the shelled barnacles, and it is possible to trace intermediate evolutionary stages in the development of parasitism in the coral and shark barnacles.

Most of the coral barnacles (Chapter 4) are epizoic on stony corals, but they feed on plankton like the rocky shore species. However, in the process of sweeping the water with their net of

The coral-eating barnacle *Hoekia monticulariae* (left), compared with a filter-feeding coral barnacle *Boscia anglicum*. The six feeding appendages (cirri) have lost their original function in *Hoekia*. No. 4 is now used as an antenna. The enlarged mandibles (m) are used to bite off the coral tissue which is stimulated to grow through the aperture of the shell (dotted lines). s: attachment to scutal valves; p: penis. (Redrawn from Ross and Newman)

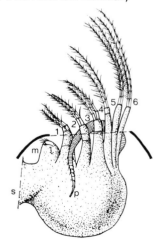

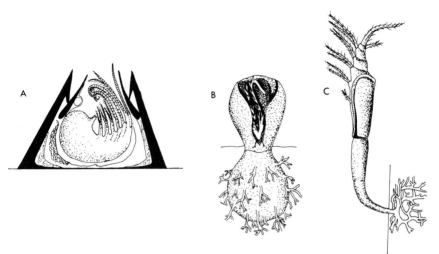

The evolution of parasitism in barnacles. (A) A rocky shore barnacle showing shell (dark), body and cirri. (B) The shark barnacle *Anelasma squalicola* has a reduced shell and attaches to its host by a system of rootlets which have assumed a feeding function. Host tissue is liquefied and absorbed by the rootlets but the barnacle still feeds in the normal way. (C) The worm barnacle *Rhizolepas annelidicola* similarly takes food from its host's tissues but is no longer a filter feeder having lost its digestive system. (From various sources)

legs they unavoidably tear away coral tissue which may be growing over their shells. This coral tissue, together with the captured plankton, is removed by another set of cirri and eaten.

One of the coral barnacles possesses a large pair of jaws as well as its feeding cirri and actively nips off and eats the coral tissue which grows over its shell. *Hoekia*, another coral barnacle, has similar large jaws and relies entirely on the coral tissue for its nutrition—its normal feeding cirri have almost completely atrophied. There is evidence that this barnacle has somehow gained a metabolic control of its coral host for coral growth in its vicinity is stimulated.

The shark barnacle, *Anelasma*, lives in a pocket of skin at the base of the dorsal spines of dogfish and is similarly half-way along the path to complete parasitism. This barnacle uses a system of roots for anchorage in the host, but at some stage these roots assumed a feeding function, for they liquify and absorb underlying host tissue. Like the semiparasitic coral barnacles, it still feeds with its cirri in the normal barnacle manner.

51

Epizoic hydroids living on the exposed parts of burrowing bivalve shells. (Intertidal, Magnetic Island)

The colonial hydrozoan *Stylactella niotha* lives only on the shell of the dog whelk *Niotha albescens*. It camouflages the snail and gains a surface for settlement on the soft mud where it lives. Epizoic hydroids may also steal their host's food. (Intertidal, Lizard Island)

Goose barnacles near the mouth of the coral crayfish may share in its food and respiratory currents. (20m. Torres Straits)

Dichelaspis, a goose barnacle which lives in the protected gills of crayfish, has reduced shell plates. (30m. off Magnetic Island)

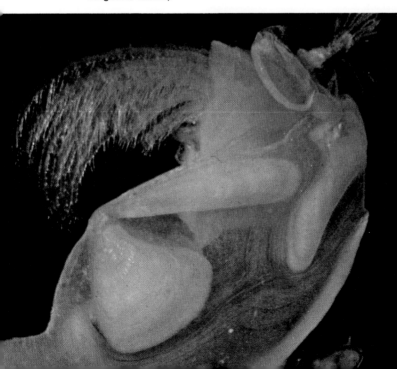

The worm barnacle *Rhizolepas* has passed this threshold into total parasitism. The roots which similarly have been used for anchorage now permeate the host, surrounding its intestine. The host's digested food is taken by the roots as it is absorbed through the wall of the intestine. *Rhizolepas* no longer uses its cirri for feeding and lacks both mouthparts and gut.

Barnacles of the order Ascothoracica parasitize echinoderms and antipatharian corals. *Syngoga*, which is an ectoparasite of crinoids, retains some crustacean features, but *Laura*, which is an ectoparasite of antipatharians, is extremely degenerate and consists of little more than a bivalved carapace embedded in the host's polyps. Dendrogastrids, which live bathed in the body fluids of sea urchins and starfishes, consist of a minute body and testes surrounded by a relatively huge branching ovary. They are imprisoned in their host's body cavities, but mature eggs are expelled through a small hole made in the host's body wall by the adult.

The order Rhizocephala contains members which are among the most degenerate of all parasites. *Sacculina*, a parasite of crabs, consists of a complex of roots (called the internal sacculina) which permeate the host's soft parts and a sac-like mass of reproductive tissue (called the external sacculina) which protrudes from the crab's abdomen.

The nauplius larva of *Sacculina* resembles those of other barnacles but lacks mouth and gut. The female cyprid larva attaches by its antennae to the bristle of a crab and then molts, discarding its legs and associated musculature. The remaining mass of cells, called the kentrogon larva, is injected through the host's body wall by a stylet. These cells then form rootlets which penetrate the host tissue, digesting and absorbing nutrients. Ultimately the entire host is invaded by the internal sacculina, which then buds off to form an ovary. The ovary enlarges and breaks through the host's soft abdomen to become the external sacculina. Male larvae are planktonic and neotenic (sexually mature); they settle on and fertilize the external sacculina.

The host crabs cease to molt and in many cases may be castrated by the parasites' invading rootlets. Female crabs cease to produce eggs, and castrated males change their secondary sex characteristics to resemble females. Commercially valuable crabs are rendered worthless by this destructive parasite. The

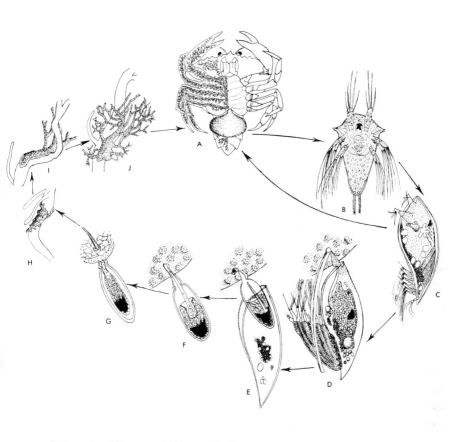

Life cycle of the parasitic barnacle *Sacculina*. (a) Mature adult in crab showing roots ('hyphae') and external ovary. (b) Nauplius larva hatched from eggs
in sea water. (c) Nauplius develops into a cypris larva and seeks its host. (The
male cypris larva settles directly on the female and fertilizes her). (d) Female
cypris settles at the base of a spine of a newly molted crab and casts off her
legs. (e) The envelope of the cypris is cast off and the kentrogon stage
begins. (f) Kentrogon forms a dart and, (g) a mass of cells is injected into the
host. (h, i, and j) The rootlet system invades the host, finally forming a bud of
reproductive tissue which becomes the external sacculina. (From various
sources)

Certain animals follow others as they feed taking scraps or food not eaten by them. At night these cardinalfishes follow stingrays into the shallows and eat animals uncovered by them. (2 m. Lizard Island)

The commensal emperor shrimp *Periclimenes imperator* feeds on its host's mucus and on materials it can reach from its nudibranch host *Ceratosoma cornigerum*. (10m. Palm Island.)

Corals and giant clams *(Tridacna)*, thanks to their mutualistic algae (Zooxanthellae), are essentially primary producers. (3m. Lizard Island)

The pear-shaped solitary coral *Heteropsammia michelinii* gains mobility from
a sipunculid worm which lives in its base. The worm in turn gains protection.
(17m. Lizard Island)

Rhizocephalan *Peltogaster* similarly parasitizes hermit crabs. *Thompsonia* is highly gregarious, with as many as one hundred individuals infesting a single host. It parasitizes crabs and shrimps. Individuals of *Thompsonia* have a common network of rootlets suggesting that the external ovaries may arise by budding from a single individual. The external sacs are lost when the host molts, but a fresh generation of sacs grows from the roots, rather like the successive crops of mushrooms which develop from mycelia.

These barnacles represent the ultimate in degeneration. They have been led by parasitism to a morphology and way of life akin to those simplest of animals, the sponges and coelenterates. Indeed, they are morphologically and functionally quite similar to the lowly fungi, which likewise consist only of a mass of invading roots (hyphae) and reproductive tissue.

PARASITIC GASTROPODS

Very few of the approximately 60,000 species of snails—about 20 genera of prosobranchs, only one hundred or so species—are parasitic. Like the crustaceans, gastropods provide us with an interesting graded series ranging from little-modified micropredators and ectoparasites to grossly degenerate endoparasites unrecognizable as snails—or anything in particular. One of these degenerate snails was even thought for many years to be a part of its sea cucumber host.

Typical of the least specialized parasites are the pyramidellids which suck their host's body fluids through long proboscises. *Odostomia* species around the world feed on bivalves such as tellins, pearl oysters, mussels and scallops by perforating their mantles and taking fluids in much the same way that mosquitoes suck their terrestrial hosts' blood. Like the mosquitoes they could be classified as micropredators rather than ectoparasites. Their sole degeneration is the loss of the radula, the characteristic rasp-like feeding organ of the gastropods.

The truly parasitic gastropods belong to the families Capulidae, Eulimidae, Entoconchidae and Paedophoropodidae. All of these parasitize echinoderms, mainly the sea cucumbers.

The Capulidae, of which *Thyca* is the sole parasitic representative, is the least specialized. It is ectoparasitic, with a thin, slight-

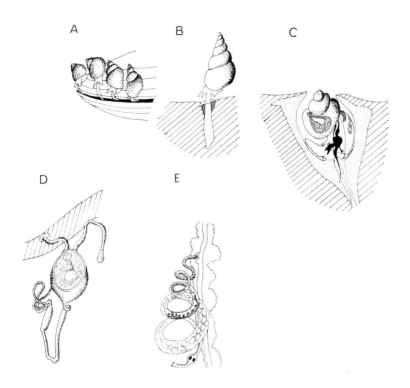

Specializations of parasitic gastropods. (A) The little-modified pyramidellid *Odostomia*, a micropredator of worms and bivalves. (B) Ectoparasitic *Mucronalia* showing the long proboscis and pseudopallium (cross hatched) in host. (C) Ectoparasitic *Stylifer* in burrow in body wall of a starfish. The snail is embedded in a large pseudopallium (stippled). (D) Ectoparasitic *Gasterosiphon* attached to the body wall and communicating to the exterior via a canal. The pseudopallium is the central sac. (E) Endoparasitic *Entoconcha* attached to the ventral vessel on the gut of a sea cucumber. It is highly degenerate, consisting of a tube filled with ovary (stippled) and testes (dark). (Redrawn from Caullery and others)

ly coiled shell and a long proboscis which it thrusts deep into its starfish host's body cavity. Like the other parasitic gastropods, it lacks a radula.

The less specialized eulimids such as *Eulima* are similar to *Thyca*, but others such as *Pelseneeria* are more specialized for their ectoparasitic life. They have thin shells, vestigial locomotory feet, are hermaphroditic and have a newly-evolved organ, the pseudopallium, which partially surrounds them.

Luminous bacteria cultured on each side of the lower jaw of the pinecone fish *Cleidopus* may be used to attract prey. Photo by Rudie Kuiter. (Sydney)

Cleanerfishes are key organisms in the fish community for they remove parasites from other fishes. A pair of blue streak cleanerfish *Labroides dimidiatus* groom a large grouper *Epinephelus tauvina*. The blue fish just above the grouper are *Assessor macneilli*. Photo by Walt Deas. (Heron Island)

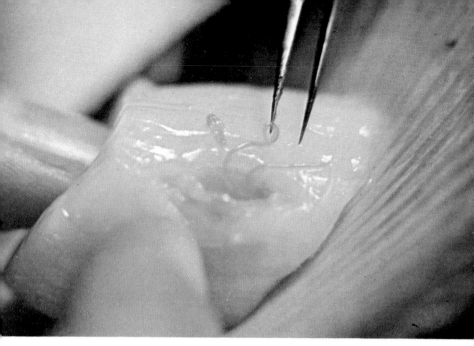

Fourth stage larva of the nematode *Sulcascaris sulcata* in the adductor muscle of the scallop *Amusium balloti.* The adult is a large ascaricoid nematode in the gut of turtles. Photo by L. Cannon.

Third stage larva of the nematode *Anisakis* on the gut of the little tuna. If eaten by dolphins these worms mature in their stomachs; if eaten by man they may cause the disease eosinophilic granulatomata. Photo by L. Cannon.

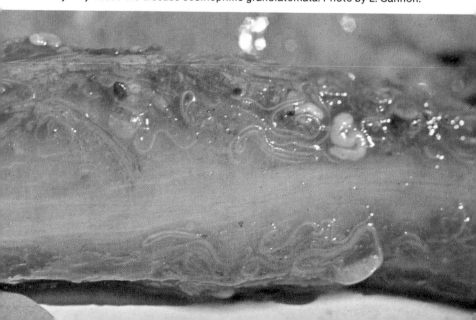

The eulimids *Mucronalia* and *Stylifer*, which live permanently embedded in the body walls of sea cucumbers, show greater degeneration. They have reduced, blind-ending alimentary canals and neither salivary glands, stomachs, nor livers.

Gasterosiphon is more degenerate again. Although it lives inside sea cucumbers, it is actually an ectoparasite for it still maintains a communication with the exterior. Its foot is rudimentary and it has no eyes, tentacles, shell, heart or gut. Its large proboscis is permanently attached to one of its host's blood vessels. The pseudopallium is enormously expanded, completely enveloping the entire body.

The most degenerate of the eulimids, *Diacolax*, also lives inside sea cucumbers and consists only of a short digestive diverticulum and an ovary enveloped by a large pseudopallium.

The family Entoconchidae superficially resembles *Diacolax* but spans all the stages between ecto- and endo-parasitism. They have lost all sensory organs, the nervous system, the visceral mass, shell and circulatory system. Their reproductive systems are extremely specialized—the ovaries are hypertrophied and the pseudopallium is enlarged and functions as a brood chamber. Males are so degenerate that they consist of only a vas deferens and a single testis embedded in the oviduct of the female.

Enteroxenos and *Thyonicola*, which are also endoparasites of sea cucumbers, represent the ultimate degeneration possible in an organism. The females consist of only a single ovary in a brood chamber which is a simple cavity in the host's tissues. The neotenic male, reduced to a single testis, lives implanted in the single ovary, the female. This surely is reducing the living organism to its fundamentals!

Chapter 3. Adaptations of Symbionts

The sea holds innumerable examples of symbiosis involving many different and often unlikely partners. These range from the opportunistic jacks, which will associate with a jellyfish or shark if one is present, to the extremely specialized endoparasitic snails—virtually only an ovary and a testis—which live bathed in their hosts' body fluids.

Although each association has its own unique character all have certain basic similarities, for similar problems confront most symbionts and often similar solutions to them have evolved.

Some of the problems facing a symbiont and their solutions may be summarized thus:

PROBLEM	SOLUTION
Location and recognition of host:	specialized larval behavior and life cycles.
High wastage of young:	increase in fecundity; evolution of hermaphroditism and neotenic males.
Penetration of host defenses:	weapons of penetration; physiological and biochemical 'disguises'.
A life bound to the biology of another organism:	atrophy, hypertrophy and change in function of organs; physiological and behavioral adaptations.

Parasitic cymothoid isopod sucking blood and body fluids from the tail of the bullseye *Priacanthus cruentatus.* Photo by Roger Steene. (Pixie Reef)

A large female isopod of the genus *Anilocra* that has settled between the eyes of a black bar soldierfish, *Myripristis jacobus.* Males may be found elsewhere on the fish. Photo by Dr. Patrick L. Colin. (15m. Mona Island, Puerto Rico)

Parasitic isopods attacking a lanternfish of the family Myctophidae. Photo taken near the surface at night by Charles Arneson. (Carbinero, Puerto Rico)

The bodies of the great whales are exploited by a range of animals seeking shelter, a surface for settlement, and food. Whale "lice", the isopod *Cyamus scammoni* and whale barnacles, *Cryptolepas rhachianecti*, from the gray whale *Eschrichtius gibbosus.* Photo by Ken Lucas, Steinhart Aquarium.

Exploitation of host without damaging it:	reduction in size; use of renewable resources; energetics balance of host and symbiont; population regulation by competition.
Competition for limited resources:	territoriality; intra- and interspecific aggression.

HOST LOCATION

The larvae and young of most marine organisms must find a suitable environment where they can grow, develop into adults and ultimately reproduce. And they often must do this against the most prohibitive of odds! During their lifetimes even the most common rocky shore barnacles must produce tens of millions of eggs for every one or two adults which successfully reach maturity. How much more difficult must it be for the very specialized sea snake barnacle to find its host in the vastness of the open sea?

Most larvae are dispersed in the plankton where they develop through a number of stages, the final one being a specialized searching stage. These larvae use several different mechanisms for locating a specific habitat or host.

Often the searching larva undergoes an alteration in its reaction to light or gravity which will place it in the general vicinity of its goal. A larval fish which will live on the sea floor might become negatively phototropic and move to the bottom, while one which lives among drifting seaweeds might develop a strongly positive phototropism and move to the surface.

Those larvae that can actively swim may have a sensitivity to a specific chemical which they follow to its source. It is known that the messmate *Carapus* and the scaleworm *Arctonoe* swim along until they encounter a chemical contained in their prospective sea cucumber host's mucus and then follow the chemical to its source.

The reactions of the freshwater mite *Unionicola* may be similar to those of some marine symbionts. The mite develops in the plankton until it becomes mature, when it develops a negative phototropism which takes it to the muddy bottom. It then swims close to the bottom until it encounters and recognizes an exudate

The messmate *Carapus bermudensis,* when a larva, finds its host by locating and following a chemical specific to the sea cucumber's mucus. Photo by Dr. Walter A. Starck II.

from its bivalve host (chemotaxis) and then follows a concentration gradient to its host (rheotaxis). The sea is similarly full of these invisible trails, some probably meaningful to searching larvae.

It is difficult to envisage how the microscopic larvae of most invertebrates could follow such chemical trails, for they are at the mercy of the waves and currents. Host location may be affected by other mechanisms—for example, by a remarkable 'molecular touch' sense possessed by the cyprid larvae of barnacles and probably by the larvae of some other invertebrates. Cyprid larvae can detect minute amounts, even single molecular layers, of a protein known as arthropodins. Arthropodins is produced in the new cuti-

The head and thorax of some of the cymothoid copepods have invaded their host's bodies. The egg sacs and abdomen of this species are external but the head is buried in the host's heart. Photo by L. Cannon.

The cymothoid copepods are very degenerate, lacking appendages, and have simplified digestive tracts. Note the parasite trailing from the anal fin of the batfish *Platax*. (Magnetic Island).

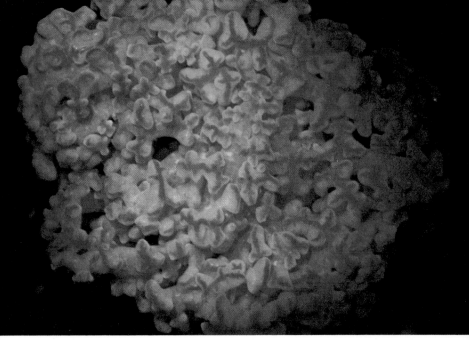

Among the most degenerate of the parasites are the ascothoracid barnacles. This dendrogastrid lives bathed in the body fluids of starfishes. It consists of only a minute body and a very large ovary. (Lizard Island)

The dendrogastrid would be unidentifiable as a barnacle or crustacean but for its larvae. Shown here are the eggs and newly hatched nauplius larvae.

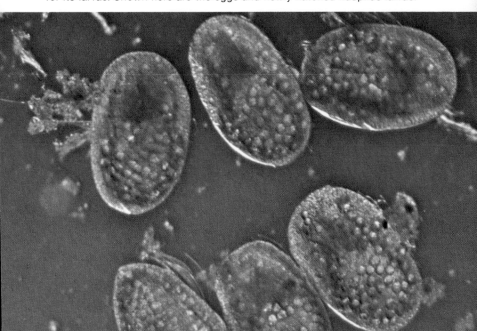

cle of barnacles and other arthropods and is chemically modified by tanning to form sclerotin, the hard material of cuticle. The molecular touch sense organ is located on the end of the cyprid's antennule and, as the larva moves around, it constantly 'feels' the molecular structure of its substrate with its antennule. By locating arthropodins molecules the larva ensures that it is in the immediate vicinity of established barnacles and that the region must therefore be suitable for settlement. The resulting gregarious distribution also ensures that the sessile adults will be close to other barnacles, thereby allowing cross-fertilization.

The molecular touch mechanism is thought to be similar to the lock-and-key effect of the antigen-antibody interaction. Experiments with extracts from a range of organisms from algae to fish show that barnacle larvae can differentiate between the arthropodins of their own species, those of other barnacles and those from other crustaceans.

Although the above refers to the non-symbiotic rocky shore barnacles, it is quite possible that the larvae of symbiotic barnacles use similar mechanisms to locate their hosts. For example, the larvae of *Sacculina* tend to settle on newly molted crabs which would be rich in arthropodins. Even sea snake, turtle and cetacean barnacles might recognize keratin, the protein of skin, which is not unlike arthropodins. Settling on a newly molted or sloughed body surface would also ensure that they remained on that cuticle or skin for the longest possible period before it too is shed.

A sensitivity to arthropodins might be a predisposition towards epizoism in barnacles, for cyprids might accidentally settle on other crustaceans that produce arthropodins. With a mutation in the code of the lock-and-key mechanism, the cyprids might also have been able to recognize the body surfaces of marine organisms other than crustaceans.

Although direct evidence of methods of host location is lacking, one can reasonably infer that similar specific responses to light, gravity, current, chemical scents and molecular structure have evolved among the symbionts to facilitate host location. Whatever the mechanisms used by the larvae to locate their hosts, the odds of any individual succeeding must be heavily weighed against it. Accordingly, it is a characteristic of symbionts that huge numbers of young must be produced.

(Above and Below) Larvae of the small sea snake barnacle *Platylepas ophiophilus,* pictured below on the tail of the sea snake *Aipysurus laevis* (head shown above), must locate their hosts in the vast area of the ocean. (off Magnetic Island)

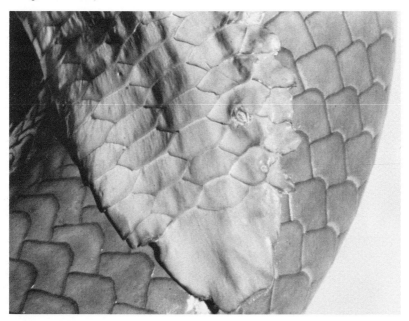

71

The sand crab *Portunus pelagicus* with the reproductive sac of *Sacculina granifera* protruding from its abdomen. Photo by L. Cannon.

The ascothoracid barnacle *Sacculina* consists only of internal rootlets or 'hyphae' and an external reproductive bag seen here protruding from the crab's abdomen. (20m. Keeper Reef)

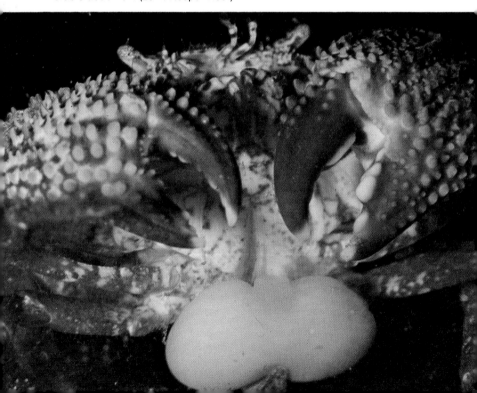

Thompsonia is a close relative of the parasitic barnacle *Sacculina*. (Above) Its numerous egg sacs form clusters on the host's body. (Below) Detail of the reproductive sacs. (100m. Coil Reef)

INCREASED PRODUCTION OF YOUNG, SPECIALIZATIONS IN REPRODUCTION

The inordinate importance of an increase in egg production is well illustrated by those extremely degenerate endoparasites which are almost entirely ovary. This high fecundity is very well documented, especially in the helminth worms which produce tens of thousands of times more eggs than their free-living counterparts.

A very basic adaptation to symbiosis is often a change in the body size of the female, for a larger organism can produce more eggs of a similar size than a small one. Many of the female parasitic nematode worms are therefore much larger than free-living ones. There is a corresponding decrease in the size of the male, for the resources he takes may be more gainfully used by the female. The evolution of dwarf males, not much more than testes, is the culmination of this trend.

The eggs of symbionts are generally very small and not heavily yolked so that larger numbers can be produced. This means that the larvae are almost invariably minute and must either develop in the plankton or find an intermediate host to develop in.

However, not all symbionts have a higher fecundity than their free-living relatives. Several symbiotic barnacles I have examined produced no more larvae than equivalent rocky shore species, but that number was, in any case, enormous. Perhaps those barnacles are physiologically not capable of a greater fecundity. Some coral shrimps do not produce more eggs in each brood than their free-living cousins, but they tend to have continuous broods while the others have seasonal broods.

The parasitic snails of sea cucumbers are not unusually fecund, possibly because they have a more direct life cycle and their sluggish hosts may be relatively easily located. Needless to say, these highly specialized parasites must also be highly efficient in host location.

A second trend in the reproduction of symbionts is the evolution of mechanisms to ensure the successful fertilization of the eggs. Many symbionts occur in low densities and there is a danger of the highly fecund female not being able to find a mate. Therefore some symbionts, such as the clam shrimps, possibly form life-long male and female bonds. Some of the endoparasites have become hermaphrodites (any two individuals can therefore mate) and others have evolved dwarf or neotenic males which live permanently attached to the females.

In addition, the reproductive cycle of the symbionts is often closely coupled to the reproductive cycle of the hosts or to behavioral cycles such as migration. Thus some of the fishes which accompany jellyfish will breed at a time that enables the juveniles to locate their seasonally-appearing hosts.

PENETRATION OF HOST DEFENSES

In many cases the hosts have defenses to keep other organisms, particularly the sessile ones, from settling on them and to keep parasites from invading their tissues. The symbiont larvae must possess ways of overcoming these defenses, which range from behavioral and physical mechanisms to cellular and immunological ones.

Most marine organisms have protective skins, cuticles or limy shells, so symbionts have evolved sharp stylets, pincers and other devices to penetrate them. Similarly, the sea urchins and starfishes

An isopod, *Ourozeuktes owenii,* has penetrated the tough defensive hide of this filefish *Meuschenia hippocrepis.* Photo by Dr. U. Erich Friese. (Sydney Harbor)

An ectoparasitic snail sucking the body fluids of the starfish *Stellaster*. Note the proboscis inserted in the host's body cavity. (10m. Magnetic Island)

Stylifer, a parasitic snail embedded in the starfish *Tamaria*. (Lizard Island)

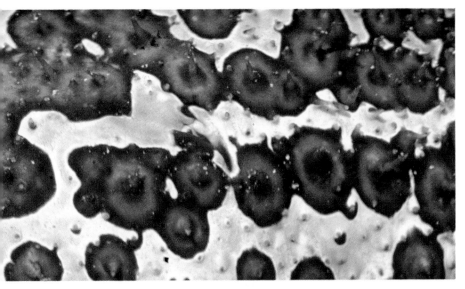

Mucronalia embedded in the sea cucumber *Bohadschia argus.* The slit at the center is the snail's communication with the exterior. (Tijou Reef)

The sea cucumber *Bohadschia* dissected to show *Mucronalia* in its body wall.

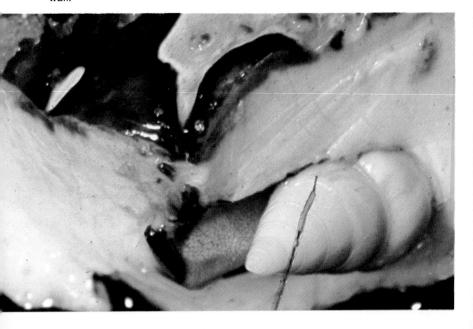

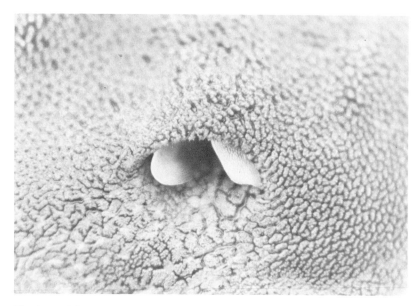

The isopod *Ourozeuktes owenii* with its tail protruding from a pouch in the abdomen of its filefish host. Photo by Dr. U. Erich Friese.

The skin of the filefish cut away to expose the isopod in its pouch in the skin of the filefish. Photo by U. Erich Friese.

have protective spines and small pincers to keep predators and fouling organisms away, but many symbionts have successfully evaded these defenses.

The marine reptiles, particularly the sea snakes, shed their skins more often than their terrestrial counterparts, and this may be an antifouling mechanism to rid themselves of sessile algae and invertebrates. The yellowbellied sea snake, a pelagic species prone to fouling, has a unique knotting behavior which is believed to be an antifouling mechanism. If something touches the skin, the snake ties itself into a slip knot and sends the knot along its body, thus wiping off the object. The leaping behavior of rays and cetaceans is also thought to be an attempt to dislodge symbionts.

Many symbionts live on corals and anemones despite their venomous stinging cells, but the ways they avoid being stung are generally not known. However, it has been recently discovered that the anemonefishes chemically disguise themselves in the host's own mucus. Other symbionts may also 'trick' their hosts in this way.

If a symbiont succeeds in invading the host's tissues it may evoke a tissue reaction such as the formation of callous tissue around it or an immunological response. Some hosts may have an inherent immunity to a particular symbiont, while others may acquire a resistance following an infection.

Most fishes have an inherent immunity to the minute monogenean flukes. These worms are blood-sucking ectoparasites of the skin and gills, but the mucus of most fishes contains antibodies which kill them. The worms have a high degree of host specificity—sometimes one species of monogenean can live on only one species of fish—because they have only developed ways of combating the specific antibodies produced by a particular host. However, one species of *Benedenia* parasitizes at least 57 species of fish because it occurs only on the hosts' corneas, a poorly vascularized area that does not produce an immune reaction.

ADAPTATIONS OF SYMBIONTS: EVOLUTION

Symbionts exhibit a multitude of morphological, physiological and behavioral adaptations to life in and on other living organisms. These changes are epitomized in the endoparasites dis-

Although attached as adults, barnacles have planktonic larvae which aid in their dispersal. These larvae follow minute chemical trails to their destination. Shown here are adult barnacles *Balanus cariosus.* Photo by Ken Lucas (Moss Beach, San Mateo County, California)

The female coral gall shrimp *Paratypton siebenrocki* produces many more eggs than related free-living shrimps. (3m. Wheeler Reef)

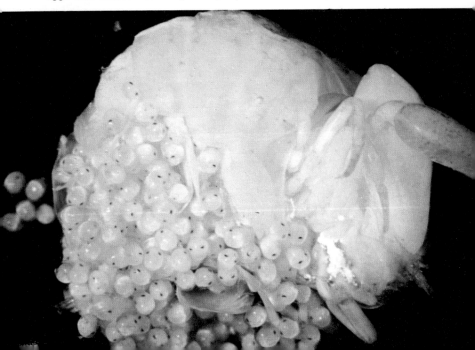

The anemonefish chemically disguises itself in the host's own mucus enabling it to live safely among the deadly tentacles. (Lodestone Reef)

The beautiful shrimp *Periclimenes colemani* lives in equilibrium with its host, the soft-bodied sea urchin *Areosoma*. (15m. Broadhurst Reef).

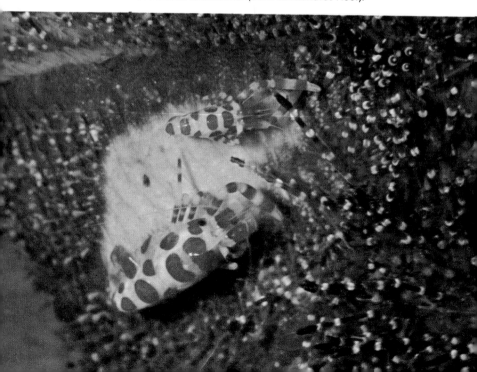

cussed in the previous chapter, but all other symbionts show their own unique adaptations to their symbiotic ways of life. The specific adaptations of various symbionts are discussed in the following chapters of this book.

The spectrum of changes ranges from minor alterations in the schooling behavior in those fishes which associate with jellyfishes through to the gross atrophications of the endoparasites.

Symbioses must have evolved and been refined over innumerable generations. Many are very ancient—even the oldest fossil crinoids bear parasitic gastropods and myzostomid worms. Fossil barnacles are found on long-extinct sea turtles and corals, and pea crabs have been found in fossilized bivalves. The sea has been buffered from the major climatic changes of the ice ages, and many of its inhabitants have changed little since the Palaeozoic.

The rate of evolution of symbionts, particularly the parasites, has probably been much slower than that of their hosts. The host provides part or all of the symbiont's environment and effectively buffers it from changes in the external physical and biotic environments. Because of their specialized ecological niches the symbionts also tend to escape from the pressures of interspecific competition. However, a complete reliance on another animal can be dangerous, for if it becomes reduced in numbers or extinct the symbionts also will become extinct.

The evolution of symbionts must also have paralleled that of their hosts. If the hosts became ecologically isolated and subsequently speciated, so too might their symbionts have speciated in time. The mobility of the hosts must also have exerted a powerful selective influence on symbionts, particularly on their life cycles. No doubt this would be greatly complicated if two or more hosts were involved.

RESOURCE UTILIZATION: ENERGETICS OF SYMBIOSIS

The symbionts other than the commensals and parasites obtain shelter, protection, a surface for settlement, feeding and respiratory currents or transport without physically taking anything from the host. The commensals which steal scraps of their host's food or eat its mucus physically remove something from the host even though they might not do any harm. The parasites which suck their host's blood, eat its tissues or steal digested food are obviously taking materials of importance to the host. Often the host may be able to sustain this loss without harm, but sometimes its condition may suffer and vital functions such as reproduction might be affected. In extreme cases the host may die. It is obviously in a symbiont's best interest not to kill the host before its own life cycle has concluded, thus a 'good' parasite will harm its host little.

The food a commensal or parasite takes from its host and the harm incurred by the host are difficult to assess. To date only one detailed study has been conducted on the energy budget of a marine commensal/parasitic association. The symbiosis studied involved the crab *Echinoecus pentagonus* and its host, the long-spined urchin *Echinothrix calamaris*. The males of this crab live around the host's anal area, the peristome, while the female lives encased in a gall within the urchin's rectum. The smaller males can enter this gall but the females cannot leave.

Energy flow model of the relationships between the urchin crab *Echinoecus pentagonus* and the long-spined sea urchin *Echinothrix calamaris*. The energy value of the host's food (input into the model) is shown at the left. The energy lost undigested in the feces and in respiration are indicated. The remaining energy is available for storage, growth and excretion. The energy taken by the crabs is indicated on the right. (Redrawn from Castro)

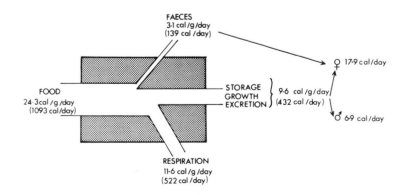

Many commensals, such as this xanthid crab, feed on the mucus of stony corals. (3m. Lodestone Reef)

The large pincer of the male (smaller one) pearl oyster shrimp *Conchodytes meleagrinae* may be used to kill or drive away rivals. (5m. Lodestone Reef)

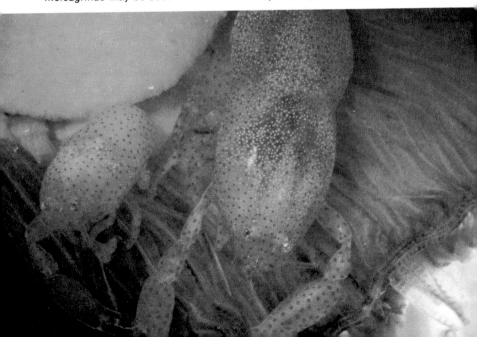

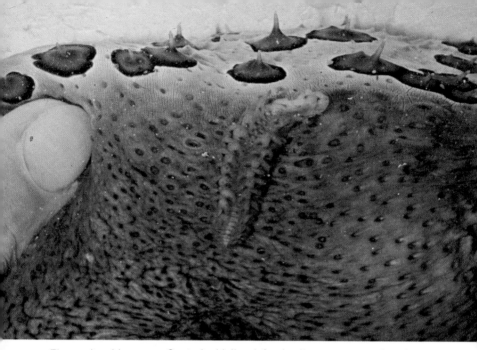

The polynoid worm *Gastrolepidella* defends its sea cucumber host *Bohadschia argus* from other worms. Usually only one worm lives on each host. (3m. Lodestone Reef)

Symbionts are invariably tied to the morphology and general biology of their host. A majid crab is perfectly camouflaged here on its soft coral host. (20m. Lodestone Reef)

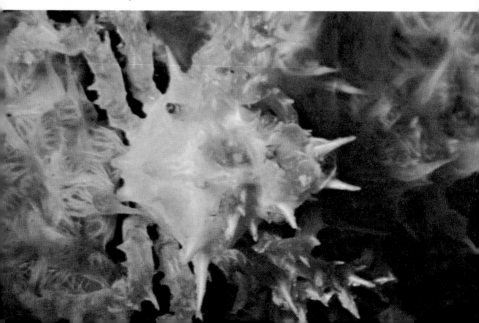

The female feeds on the host's blood cells (coelomocytes) and undigested food in the urchin's feces. It uses special appendages to sort through the feces to extract algae, sponge tissue, living bacteria, ciliates and nematodes. The male crab eats its host's epithelial tissue and tube feet.

The energy or calorific value of the food consumed each day by the host was calculated and the calorific value of the feces and the energy lost in respiration were deducted from this. The resultant energy is available for storage, growth and excretion and for the use of the symbiotic crabs. The amount of energy the crabs took (25 cal./day) was only a fraction of that lost in the feces (139 cal./day) or that available for storage, growth and excretion (432 cal./day). The rate of ingestion of host tissue (male crab 4 mg./day; female crab 10 mg./day) proceeded at a rate similar to its regeneration, so the association remained in equilibrium.

Although no such studies have been conducted on other marine symbioses, it has recently been found that the mucus secreted by corals is wax-rich, with a high energy value, and is an important medium in the recycling of energy in coral reefs. Mucus produced by the soft-bodied coelenterates and molluscs attracts many commensals which can feed on this easily renewable material without harming the host.

COMPETITION AMONG SYMBIONTS

The energetics of the sea urchin/crab association suggests that one pair of crabs can live in harmony with the host but that two pairs might not. It is therefore common to find that many symbionts regulate their own numbers by territorial behavior or by excluding or subjugating newcomers.

Thus the messmate *Carapus* eats the larvae of its own species if they invade its sea cucumber host. Established pairs of anemonefish (*Amphiprion*) take less drastic measures: they either drive away juveniles seeking a host or subjugate them, preventing them from maturing until one of the established adults dies.

Several of the shrimps which inhabit bivalves live in pairs, a large female with a smaller male. The latter has a large claw of unknown function, and it has been suggested that the male uses this claw to dispatch any newcomers. Even some false cowry shells

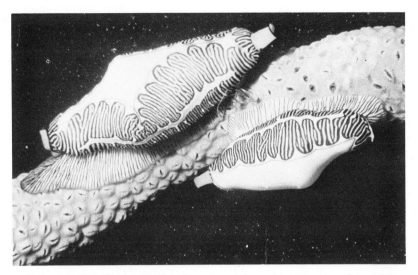

False cowries *Cyphoma signatum* on a plexaurid gorgonian. Some false cowries are territorial and may have an encounter over the position of their boundaries. Photo by David L. Ballantine.

living on gorgonians are territorial and have disputes over their boundaries.

The sea cucumber scaleworm *Arctonoe* is strongly territorial and drives away any newcomers seeking to establish themselves on its host. In areas of high host densities there are territorial disputes between neighbors and some worms may be displaced from their hosts. These worms may become free-living until they locate suitable hosts.

Arctonoe is strongly territorial because it needs to shelter in its host's mouth—and as the host only has one mouth, only one worm can live on each host.

The resources available to a symbiont are very limited compared with those available to most aposymbiotic or free-living organisms. If the limited resources supplied by the host are seriously over-taxed, its health may suffer and possibly less food will be available to the symbiont. And if the host dies, so might the symbiont. There are thus very strong pressures of natural selection to prevent symbionts from over-exploiting their hosts: the genes of symbionts which kill their hosts would be rapidly selected out of the population.

The coral reef ecosystem, the most complex of all ecosystems, is based on an animal/plant symbiosis. Many thousands of species live in close proximity and many symbiotic partnerships have evolved. (5m. Lizard Island)

Cliona, the boring sponge, bioerodes living and dead corals. This is a broken section of living *Pocillopora* moderately damaged. (4m. Palm Island)

Serpulid polychaete worms may crowd encrusting and massive corals. (7m. Magnetic Island)

A conical white operculum blocks the burrow of the serpulid when it retreats. (7m. Magnetic Island)

CONCLUSION

These are only a few of the difficulties encountered by organisms which live in intimate contact with another species. Likewise the adaptations described here are only a few of those which have been evolved by the symbiont.

Much remains to be learned about the biology of symbiosis. To date studies have generally been piecemeal and often do not follow any pattern. It would be enlightening, for instance, to begin to quantify the relationships between the symbiont and its host. How much energy, in terms of calories, is contained in mucus? How much energy can a particular host afford to lose? How much extra energy must it utilize to make up for that lost to the symbiont? What proportion of a symbiont's energy goes into reproduction and how does this compare with that of a similar free-living organism? The questions are endless.

PART II: SYMBIOSIS IN MARINE ECOSYSTEMS
Chapter 4. Coral Reefs

The coral reef is the most beautiful and mysterious ecosystem in the biosphere. A visit to a reef leaves one with a lasting dream of petrified gardens and fairyland inhabitants, and an awe of the humble polyps which construct underwater mountains.

THE CORAL REEF COMMUNITY

A geologist, P.E. Cloud, described the coral reef community as *"essentially a steady state community of high population density, intense calcium metabolism and complex nutrient recycling, generally surrounded by waters of relatively low nutrient and plankton content."*

Its high population density and species diversity are unique. One can only guess how many species might be represented on, for example, an average patch reef of the Great Barrier Reef. The number would be tens of thousands, perhaps nearing one hundred thousand species. There might be 1,000-1,500 species of fishes alone, and these may harbor up to 50,000 species of parasites. This figure might be doubled on a reef in the Philippines.

The results of the intense calcium metabolism are obvious. Biological activity has filled vast sections of tropical seas and created coral islands and even mountain ranges. At Enewetak Atoll in the Pacific the coral limestone is over one kilometer in thickness. Along the edge of the continental shelf of northeastern Australia corals have produced a broken rampart of over 2,500 separate reefs which stretch 2,000 kilometers. The sea which washes onto this rampart, the Great Barrier Reef, is aptly called the Coral Sea.

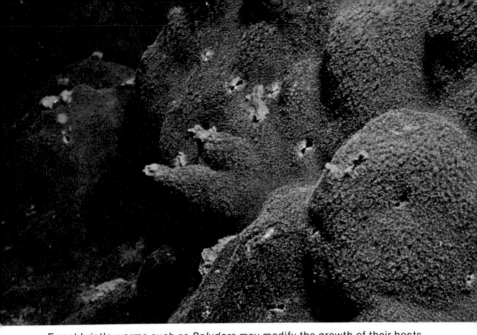

Errant bristle worms such as *Polydora* may modify the growth of their hosts. (5m. Magnetic Island)

Camouflaged scale worm or polynoid on the coral *Pocillopora*. (3m. Lodestone Reef)

The exquisite peacock worm *Spirobranchus giganteus* (above and below) lives in the massive coral *Porites.* The plumes are used for feeding and respiration. (3m. Lizard Island)

Many small reef fishes shelter and sleep in the branching corals. (2m. Prawn Reef)

Little is known of the nutrient cycles of a coral reef. In the past a reef was thought to be an animal-dominated or heterotrophic system bathed in and sustained by a rich plankton soup. It is now recognized that a coral reef is actually a plant-dominated or autotrophic system: it is maintained by the sun's radiant energy being transformed by the photosynthetic organisms of the community. A system of complex nutrient recycling enables the superlatively co-adapted reef species to live in the relatively barren surrounding waters. The coral reef is an oasis.

The co-adaptations of the coral reef species are epitomized in the inordinately high numbers of symbiotic associations found there. Indeed, the very existence of coral reefs depends on a mutualistic symbiosis between an animal, the coral polyp, and a plant, the zooxanthellae. (This complex association is the subject of a later chapter).

Symbiotic associations greatly swell the number of species on coral reefs; probably every species has its complement of symbionts, mainly parasites. It has been estimated that the 1,000 fish species of Heron Island on the Great Barrier Reef host 2,000 species of ectoparasitic monogenean flukes alone, and the total number of ectoparasites of these fishes may be 20,000 species. Molluscs are heavily parasitized by intermediate stages of fish flukes—thirty different species of flukes have been found in one single species of snail.

Some groups that are exclusively free-living (aposymbiotic) in temperate seas are markedly symbiotic in the coral reef community. For example, over half the shrimps collected from an Indian Ocean coral reef were symbionts on a variety of other animals. Likewise most of the barnacles in Indo-Pacific coral reefs are symbionts of corals.

Many of the partnerships involving the coral reef, molluscs, ascidians, echinoderms and fishes are discussed in later chapters. In this chapter the associations directly involving the corals are described and the general implications of the phenomenon of symbiosis in the coral reef ecosystem are discussed.

CORAL CLASSIFICATION

The corals are a diverse group belonging to the phylum Coelenterata, class Anthozoa. The important stony or reef-forming corals (subclass Zoantharia, order Scleractinia) have six tentacles or multiples of six and a skeleton of calcium carbonate. The black corals and sea whips (order Antipatharia) also have six tentacles but have a horny skeleton.

The soft corals (subclass Alcyonaria, order Alcyonacea) have eight tentacles or multiples of eight, and most species are soft-bodied with a framework of calcareous spicules. The related sea fans (order Gorgonacea) have a horny skeleton and a fragile upright growth form.

Organisms of every major phylum have entered into symbiotic associations with the different corals.

SYMBIONTS OF STONY CORALS
Algae

The role of the endozoic dinoflagellates, the zooxanthellae and the filamentous algae are outlined in Chapter 9.

Gall created in the coral *Pocillopora* by the crab *Hapalocarcinus marsupialis*. (3m. Lodestone Reef)

View through the top of a partly formed gall in the coral *Pocillopora*. Note the large top chamber and smaller bottom one occupied by the young crab. (3m. Lodestone Reef)

Galls of *Hapalocarcinus* in *Seriatopora*. (5m. Palm Island)

Stages in gall formation in the coral *Seriatopora hystrix*. (4m. Palm Island)

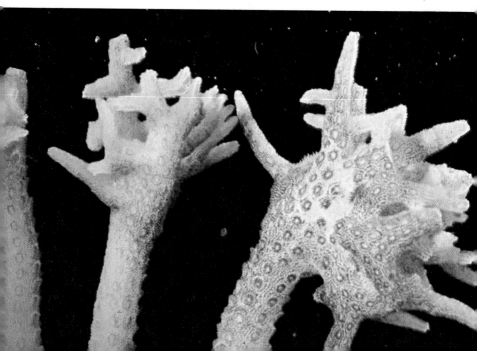

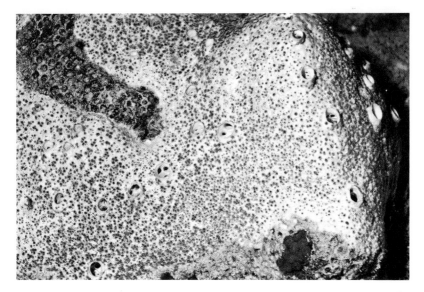

The boring sponge *Cliona delitrix* has invaded this coral head and has almost completely covered it. The sponge itself is associated with a zoanthid *Parazoanthus parasiticus*. Photo by Dr. Patrick L. Colin. (12m. Desecheo Island, Puerto Rico)

Sponges

Many species of sponges encrust the dead coral skeletons on the coral reef, but few sponges form regular associations with living corals. However, in the West Indies the sponge *Mycale laevis* regularly lives adjoining living corals of the reef slopes. It has been suggested that this association is mutually beneficial because of an exchange of metabolic products. But sponges more often have a deleterious effect on corals. Some Barrier Reef sponges compete with corals for space and may smother them, killing the polyps.

More destructive is a microscopic sponge, *Cliona*, which bores into the skeletons of living and dead corals, honeycombing them. It is a major cause of disintegration of branching corals, dead coral and coral rock, and living and dead mollusc shells.

The sponge excavates the extensive galleries and tunnels by chemical and mechanical means. Specialized mesenchyme cells or amoebocytes attach to the calcium carbonate substrate and release a secretion along the rims of their cytoplasm. The chemical etches the surface, gradually undercutting a section corresponding to the cell's dimensions (about 85 x 60 micrometers) until a chip is freed. The chip is then picked up and carried by amoeboid transport to finally be expelled via the exhalent current. The composition of the secretion is unknown, but it probably consists of an acid to dissolve the mineral and enzymes to digest the organic matrix of the coral skeleton.

It has been estimated that erosion on coral reefs occurs at a rate of about 2 kg./m.^{2}/year, of which half is due to living agencies, called 'bioerosion'. About one half of the bioerosion is caused by grazing fishes that bite off pieces of dead coral and grind them to extract the algae, but about one third is attributed to the hidden *Cliona*.

Cliona undoubtedly gains shelter by excavating galleries, but it is not known if that is the sole reason for the association. Few, if any, of the other sponges shelter from predators.

Polychaete Worms

A number of errant or mobile polychaetes such as *Hermodice* live on the surfaces of stony corals and crawl about grazing on

The spectacular peacock worm *Spirobranchus* builds a tube in various corals often altering noticeably the growth of their hosts. Photo by Dr. Herbert R. Axelrod.

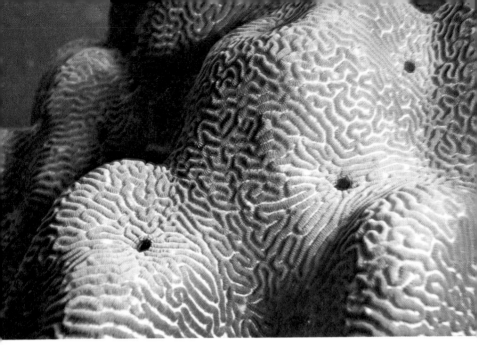

Burrows of the crab *Cryptochirus* may greatly alter the growth of their hosts, in this case *Leptoria*. (5m. Lodestone Reef)

Burrow crabs of the genus *Cryptochirus* regularly live in the solitary coral *Trachyphyllia*. (20m. Lizard Island)

Adult gall crabs have a soft carapace and produce many eggs. Compare with coral gall shrimp. (3m. Lodestone Reef)

Juvenile gall crab, one side of the coral being removed to reveal it. (4m. Palm Island)

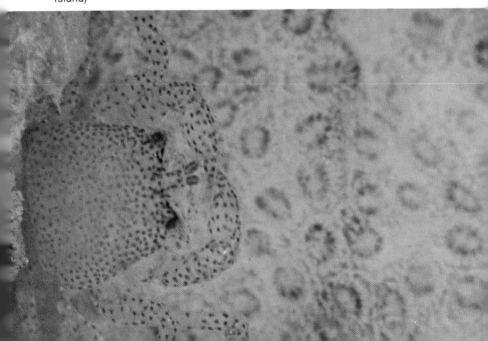

mucous secretions and portions of polyps. Several sedentary tube worms also live permanently attached to living corals, probably gaining protection from the host that envelopes the worm's tube. Best known of the tube worms is the spectacular peacock worm, *Spirobranchus*, which extends a pair of vividly colored and exquisitely spiralled feeding tentacles from its coral host. The preferred host is a massive coral, *Porites lutea.*

Polychaete worms can greatly alter the growth of their hosts. Some of the tube worms that adhere to branching corals may be covered by their hosts' skeletons and ultimately produce new branches. Similarly, the errant worm *Polydora*, which lives in a fragile membranous tube, becomes covered by the host, forming a miniature volcano. The coral's growth may be so altered by a heavy invasion of worms that it becomes unrecognizable.

Crustaceans
COPEPODS

Several small xarifid and asterocherid copepods live as commensals and parasites of corals. Some have been seen wading among the coral polyps, slashing them with specially modified razor-sharp claws. Other parasitic copepods live permanently embedded in coral tissue and have lost their appendages and exoskeletons.

DECAPODS

Larger palaemonid shrimps (*Periclimenes, Coralliocaris, Jocaste*), pistol shrimps (*Alpheus, Synalpheus*), and xanthid crabs (*Cymo, Domecia, Trapezia)* are abundant among the branching corals. In excess of forty species live in the seriatoporid and acroporid corals of the Great Barrier Reef. These vary in their host preferences, but only one shrimp species is common to both these families of corals. Strangely, the tropical Atlantic has only one species of decapod, a crab, living on its corals.

The feeding habits of few of these decapods have been investigated. It is known that the xanthid crabs cling to the coral cups with especially hooked claws and scratch other specialized brush-tipped legs back and forth among the polyp's tentacles. The mucus and other debris which adhere to the brushes are combed out by yet another set of specialized appendages and are eaten. Although some host tissue may be taken, the crab's strong territoriality limits evenly disperses the crab population, thereby ensuring that the host colony is not over-taxed.

Lybia (formerly *Melia*), a rarely encountered coral symbiont, has a unique way of feeding. It carries in its specially modified claws a pair of small sea anemones. The anemones are dragged along in the manner of mops and the detritus which adheres to them is removed and eaten. If the crab is attacked it uses the stinging anemones as weapons and thrusts them at the intruder.

Little is known of the feeding habits of the shrimps, but gut contents reveal coral mucus, detritus, expelled zooxanthellae and fragments of coral tissue. The different shrimp species live in different parts of the coral, for example among the inner branches, at the branch tips, at the base of the colony or on the polyps. Like the xanthid crabs the shrimps tend to be strongly territorial.

The coral crab *Lybia* carries anemones in each specially modified claw (below). The anemones are used to sweep up food and for defense. (Redrawn from Duerden)

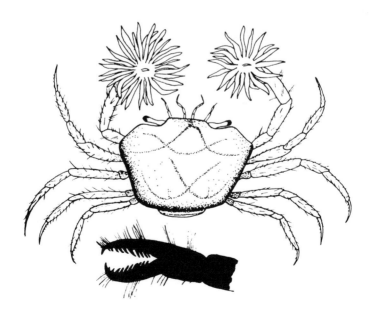

A close relative of the coral gall crab, *Cryptochirus,* makes a burrow in the large-polyped massive coral. It uses its stout carapace to block the hole. (3m. Tijou Reef)

Xanthid crabs, *Trapezia wardi* from Hawaii (above) and *T. cymodoce* from Lodestone Reef (below), as well as other species of crabs live on the branches of *Pocillopora*. Photo above by Scott Johnson; photo below by the author.

GALL CRABS

Some of the crustaceans modify their host's growth. A crab such as the Atlantic *Domecia* that habitually shelters on a given part of its host may kill the polyps beneath it and cause others to diverge in their growth to partially surround it.

This trend is highly refined in the female gall crab *Hapalocarcinus*, which lives permanently imprisoned in a gall within the Indo-Pacific seriatoporid corals but draws a feeding and respiratory current through small pores in the wall. The small, millimeter-long, post-larval female first settles on a growing tip between the buds of two forming branches. It sets up a feeding current which causes the new branches to grow around it to form a cup or chamber. As the crab grows it fills this chamber, but its feeding current is directed such that a second or larger chamber is formed by the growing coral above the first. The crab occupies this and the older one fills up with debris. The tips of the coral then converge in growth, completely enclosing the crab except for

The stages in the development of the gall of the coral gall crab *Hapalocarcinus marsupialis* in *Pocillopora*. (a) Growing tip of host. (b) Tip divides to form a branch. (c) Post-larval female settles on tip. (d) Feeding currents of young crab change   the coral's growth. (e) A second chamber forms above the first. (f) The mature crab shifts into the larger chamber and is fertilized by the free-living male. (g) The coral ultimately closes the chamber but for small holes through which the crab draws her feeding currents. The young escape through these holes. (Modified from Potts)

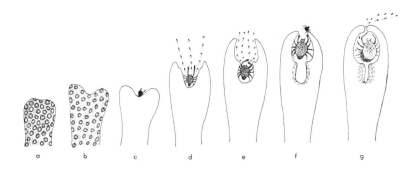

some small holes through which the current and the minute adult male can pass. The eggs are brooded by the soft-shelled female, and the newly hatched larvae pass out through the holes.

Some crabs of the same family (Hapalocarcinidae) have similarly exploited the large-polyped massive corals. The post-larval young of crabs such as *Cryptochirus* settle on a polyp and kill it. The adjoining polyps continue growing, creating a shallow pit. Ultimately, after several years, the burrow will be two to five centimeters in depth depending on the growth rate of the host. The males of most species visit the females to fertilize them, but those of *C. dimorphus* live permanently attached to the underside of the females.

A shrimp, *Paratypton*, builds a very similar gall, but only in the shelf staghorn coral *Acropora hebes*. However, unlike the gall crab both male and female live trapped together. The smaller male is relatively shrimp-like with large nippers, but the larger female is soft-shelled and bloated, without pigment and with vestigial eyes and ineffectual legs—remarkably like the female gall crab.

CORAL BARNACLES

About sixty species of barnacles belonging to the balanid or acorn barnacles live on and in the stony corals. Although barnacles are often the dominant organisms of the rocky shores there are surprisingly few on the coral reef community. It has been suggested that heavy grazing by fishes prevents their establishment, but I suspect that it may be that they cannot successfully compete for space with the faster growing colonial organisms such as corals, sponges and bryozoans which dominate the coral reef community. Rather than compete with their rivals, the coral reef barnacles have tended to exploit them in symbiosis.

The three groups of coral barnacles, represented by *Boscia* (cosmopolitan), *Ceratoconcha* (Western Atlantic) and *Cantellius* (Indo-Pacific), are thought to have evolved from different balanid stocks. They display a range of host preferences—some are widespread, others are specific. The most specialized species are the most specific in their host preferences.

The larvae of the coral barnacles settle on the coral's living tissue, somehow avoiding being stung or eaten, and grow

Ovigerous female pagurids removed from burrows in coral. Note the worm-like abdomens with hooked tails used for anchorage. An enlarged pincer is used as an operculum to block the burrow.

Like the burrowing hapalocarcinid crabs these unusual pagurids (hermit crabs) form burrows in massive corals. (5m. Tijou Reef)

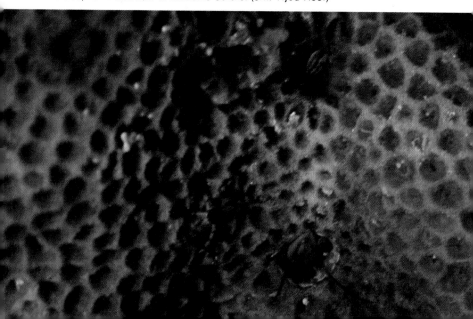

The small male of the coral gall shrimp *Paratypton* has relatively large chelae, despite its permanent imprisonment. (5m. Wheeler Reef)

Male (smaller one) and female coral gall shrimps *Paratypton siebenrocki* live imprisoned together in the shelving coral *Acropora hebes*. (Dead specimens collected at Wheeler Reef)

downward until their shells come in contact with the coral skeleton. Some of the barnacles stimulate their hosts to lay down septa on their shells, thus disguising themselves as polyps. This may not be completely successful, for I have often seen feeding scars on corals indicating that some of the wrasses have a taste for coral barnacles.

The coral barnacles feed in the manner of their rocky shore cousins but may supplement their planktonic food with fragments of their hosts. The evolution of the coral-eating barnacle *Hoekia* has been described in detail earlier.

Molluscs

Members of at least six families of gastropods and six families of bivalves are symbiotic with corals. Certain of these, the coral-boring mussels and snails, are important agents in the bioerosion of coral reefs.

The most brilliantly colored symbionts are the nudibranch gastropods. Some, like *Phestilla*, mimic the coloring and shape of their hosts, in this case the rich pink, orange or deep green *Tubastrea*. When it is on its host *Phestilla* is almost indistinguishable from it, but against another background it is very conspicuous. This nudibranch grazes on polyps and may eventually kill whole colonies; it then lays its eggs on the dead coral. Staircase shells, *Epitonium*, also feed on *Tubastrea* polyps but, unlike *Phestilla*, which has a more predatory association, they suck the host's body fluids through a long proboscis. *Epitonium* may also match the exact color of the host. Both symbionts probably extract the host's characteristic pigment and incorporate it into their own tissues for camouflage.

The muricid snails have a similar predatory or parasitic relationship with corals. The ornate *Drupa camus* feeds on the massive coral *Porites* and the branching seriatoporid corals. It produces copious amounts of mucus to prevent the stinging cells of the corals from injuring it. *Drupa* feeds by secreting salivary enzymes to liquefy the coral tissue, sucking this liquid with its long proboscis. Some sundial shells and allied cowries feed on coral tissue by grazing it with their radulas.

Almost all of the staircase shells (Epitoniidae) have a commensal, parasitic or predatory association with corals or sea anemones.

Muricid snails are predatory on or parasitize corals by secreting enzymes to liquify the coral tissue. Photo of *Drupa ricina* by Scott Johnson. (Koko Head, Hawaii)

Some parasitize the mushroom corals, *Fungia*, pinching off pieces of tentacles and tissue with their proboscises. These parasites are extremely fast-growing (a settled larva grows to maturity in two weeks) and seasonal in their occurrence.

The family Magilidae (also called Coralliophilidae), the most specialized gastropod parasites of corals, is probably descended from the less specialized muricids. *Coralliophila* and *Quoyula* are ectoparasites and live on small calcareous platforms which they secrete on their hosts, while *Leptoconchus* and *Magilopsis* excavate cavities in their host's skeleton. All the coralliophilids have a very long proboscis which they extend to reach the surrounding polyps. Like the muricids, they liquefy the coral mucus and tissue with enzymes and pump this soup up through their proboscises. They are 'good' parasites, for the tissue surrounding them is usually healthy—the host keeps pace with the damage done. Although the coral-boring coralliophilids have very thin shells, it is thought that they use them to abrade the cavities, possibly aided by an acid secretion.

A pair of coral shrimps *Hamopontonia corallicola* well camouflaged among the tentacles of the mushroom coral *Fungia actiniformis*. (15m. Palm Island)

Ovigerous pistol shrimp *Alpheus ventrosus* sheltering in the basal branches of *Pocillopora*. (3m. Lodestone Reef)

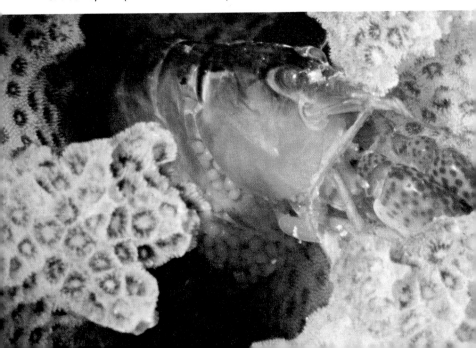

Heavy infestations of boring *Lithophaga* may weaken the strength of corals, even the massive forms. (5m. Magnetic Island)

The coral barnacle *Cantellius gregarea* and boring mussel *Lithophaga* live inside *Pocillopora*. (3m. Lodestone Reef)

Closely related to these borers is *Magilus*, which produces a very long tubular shell that is embedded in the coral host. Although its shell is elongated like that of a tube worm, certain malacologists have suggested that *Magilus* might only be a growth form of one of the other genera.

The date mussels, *Lithophaga*, are very common and contribute greatly to the bioerosion of living and dead corals, being responsible for considerable damage to standing corals. In the Red Sea, biologists have estimated they are responsible for about 20% of coral breakages (wave action is the major cause of damage). Often only a light tap is sufficient to cause a solid-looking colony to shatter for, like the marine timber borers, their extensive excavations are internal. *Lithophaga* use the coral only for shelter—they are filter-feeders and draw feeding and respiratory currents through small dumbbell-shaped apertures in the surfaces of their hosts.

Their method of boring is the subject of some conjecture. Some biologists propose that they rely on acid to dissolve the cavity, but

Boring snail *Leptoconchus* removed from its host. Note the long proboscis used for feeding on host tissue, the vestigial foot, and the reduced sensory tentacles (eyes absent). (5m. Tijou Reef)

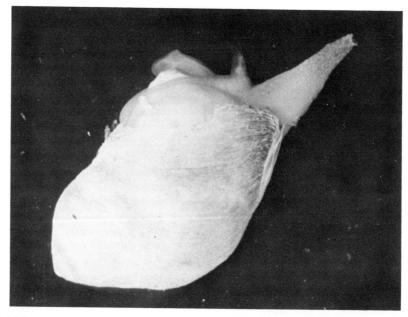

others suggest that they move their solid rasp-like shells back and forth to abrade the cavity. Although some mussels can bore through non-calcareous rock using only mechanical means, it is now thought that the coral borers do secrete an acid from their mantles and then rasp the softened rock with their shells.

Another coral mussel, *Fungiacava*, was discovered in Red Sea mushroom corals a few years ago. This bivalve has a very thin shell and excavates its cavity beneath a polyp, protruding its siphons up into the host's gut or coelenteron. It feeds on zooxanthellae expelled from the host and on waste products of digestion.

The bivalve *Pedum*, whose grinning shells lie well protected in deep fissures in the massive coral *Porites*, has been known for well over a century but biologists have been undecided on its affinities. Although it was initially placed with the pearl oysters, then the true oysters, it is now included with the scallops. *Pedum* does not excavate the deep cavity in which it lives but, like the gall crab, modifies the growth of the surrounding polyps with its feeding currents. A neighboring bivalve, the arc shell *Navicula*, similarly modifies the growth of *Porites* in this manner, but its depression is shallower and may be occupied by several individuals.

Fishes

Very many of the reef fishes have a close association with living corals. If a predator appears the damselfishes which cloud the waters of the coral reef dart down into the branching corals and remain there until it is safe to come out. At night many reef fishes shelter or sleep among these branches, well protected from the nocturnal predators. These associations are rarely very specific as almost any species of branching coral may be used for shelter. It is interesting to note that the staghorn corals are less popular than the others, possibly because their polyps have a more powerful sting.

Specific associations include the cardinalfish *Apogon leptacanthus*, which habitually schools in branching *Montipora;* the adult dominofish *Dascyllus trimaculatus*, which shelters in *Pocillopora*; and species of gobies, *Paragobiodon*, which are specific to the genera of seriatoporid corals. These gobies rarely if ever leave their hosts.

115

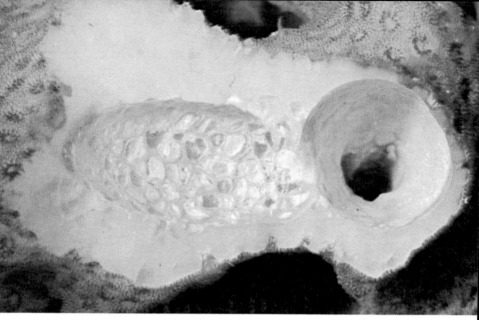

The boring mussel *Lithophaga* is important in the bioerosion of living and dead corals. This weakened branch broke when tapped. (3m. Lodestone Reef)

The inhalent and exhalent apertures of the filter-feeding *Lithophaga*. (3m. Lizard Island)

Gall on mushroom coral *Herpolitha* produced by parasitic gastropods. Note the pink regenerating tissue nearby. (3m. Lizard Island)

Boring coralliophilid snails *Leptoconchus* in the stomodeal pit of the mushroom coral *Fungia*. Endozoic algae surround the burrows. (2m. Tijou Reef)

SYMBIONTS OF ANTIPATHARIANS

The antipatharian sea whips and the valuable black corals are less common than the other corals and generally prefer deeper waters subjected to strong currents.

Sea Whip Community

The antipatharian sea whips are longer and more slender than the superficially similar gorgonian sea whips and often take a kinked or corkscrew form as they grow.

The Indo-Pacific *Cirripathes* is the host of an elegant little fish, Yonge's goby (*Cottogobius yongei*), and a number of shrimps such as *Dasycaris zanzibarica* and *Pontonides unciger*. Each of these has evolved a similar camouflage for life on the sea whip—they are almost completely transparent but for regular blotches which resemble the host's expanded polyps.

As the single branch of the host offers comparatively little protection, Yonge's goby has evolved an amusing hiding behavior. If an approaching object startles it, it will immediately jump around to the other side of the branch. If its potential enemy circles the sea whip, the goby in turn dashes back to the other side like a child hiding behind a telephone pole.

Sea whip gobies. *Cottogobius yongei* live on both gorgonian and antipatharian sea whips. Its host here is *Cirripathes anguinus*. (10m. Broadhurst Reef)

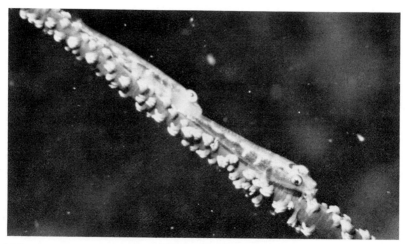

The shrimp feed mainly on mucus and organic detritus adhering to the mucus, while the goby takes plankton and organic materials that drift by. It is interesting to note that Yonge's goby is able to exploit an environment not normally available to the gobies. The gobies have evolved as specialized bottom dwellers and, being weak swimmers because of the loss of their swim bladders, the mid-water resources are beyond their reach. However, Yonge's goby can climb up sea whips using the gobies' specialized pelvic suckers and dart out to take drifting food.

The goby and shrimps are strongly territorial as each sea whip of a forest will only have between one and three fish and one or a pair of shrimps. Like the inhabitants of stony corals, they avoid overcrowding their hosts by regulating their own numbers.

Black Coral Community

The black coral trees which inhabit caves and coral faces in deep water are exploited by many symbionts. However, to date there have been few reports on black coral symbionts; the following are from my own brief observations.

Great Barrier Reef *Antipathes grandis*, a large and commercially valuable black coral, regularly hosts a sponge, small shrimps, a majid crab, a balanid barnacle, several species of brittle star (*Astrobrachion*), an oyster (*Dendrostraea*) and small fish similar to Yonge's goby.

The sponge, barnacle and oyster are all plankton feeders and must use the black coral for shelter and a substrate. The brittle stars, shrimps and crab may be commensals, feeding on the sheets of mucus and epithelial cells secreted by the host. The brittle stars have extremely long arms which they tightly coil around the branches of the black coral, looking very much like pythons coiled in a tree.

SYMBIONTS OF SOFT CORALS

The Alcyonacea, the soft rubbery corals which often dominate the shallow waters of the coral reef back and lagoon, also host a large variety of symbionts although there are few reported in the literature. Most of the following observations on three typical soft

Parasitic staircase shell *Epitonium* cf. *ulu* lays its brood of eggs on the base of its host, the mushroom coral *Fungia somervillei*. (20m. Lodestone Reef)

The similar parasitic *Epitonium* cf. *costulatum* brooding its eggs on the base of the coral *Fungia*. (20m. Broadhurst Reef)

The chrome yellow aeolid nudibranch *Phestilla melanobranchia* takes pigment from its variably colored host, *Tubastrea*. The nudibranch has eaten the coral beneath it and is laying its eggs on the dead skeleton. (3m. Magnetic Island)

Another symbiont of the chrome yellow *Tubastrea*, the staircase shell *Epitonium billeeanum*, also takes its host's pigment. (3m. Magnetic Island)

corals again come from my brief studies while researching this book and should not be considered exhaustive.

Xenia Community

The beautiful sky blue *Xenia* lives in shallow waters of Indo-Pacific reefs. It has large polyps with long feathered or pinnate tentacles which constantly move to and fro with the surge.

Xenia, like the other soft corals, has mutualistic zooxanthellae which give it its unusual color. But unlike the other soft corals its polyps lack mouths, indicating a total reliance on the photosynthetic productivity of their algae.

A small portunid crab, *Caphyra laevis*, often lives well hidden among the waving polyps. It is very well camouflaged, for the chromatophores on its carapace have 'painted' three or four of the host's polyps—tentacles and all! A small shrimp, *Hippolyte commensalis*, also lives among the polyps. It is almost completely transparent but for white and light brown bands. Little is known of the biology of these crustaceans, but it is presumed that they are commensals feeding on mucus and wastes and possibly some host tissue.

Sinularia Community

The rather large rubbery colonies of the intertidal *Sinularia* support a more diverse community of symbionts. One of the most unusual lodgers is an aberrant ctenophore or sea gooseberry, *Coeloplana*. While the vast majority of ctenophores are planktonic, *Coeloplana* is sedentary and lives beneath soft corals. At night it casts out long sticky tentacles which are carried out in the current like a fishing line. Plankton and organic material drifting by stick to these threads like flies to flypaper, and *Coeloplana* periodically pulls them in to remove the catch. The tentacles are drawn through narrow pouches in the body of the ctenophore to wipe off the food which is then carried to the mouth along ciliary paths.

Small, well camouflaged shrimps and a goby similar to Yonge's goby but smaller live on the upper surface of the soft coral. Like their cousins on the sea whips they are transparent with colored blotches which resemble their host's retracted polyps. If they are disturbed they hide beneath the soft coral's fleshy folds.

Soft corals, such as this one from the Maldive Islands, harbor a number of symbionts from various phyla. Photo by Dr. Herbert R. Axelrod.

Allied cowries and aeolid nudibranchs regularly feed on soft corals such as *Sinularia*. The large and very beautiful white egg cowry, *Ovula*, feeds on soft corals but a number of smaller species of allied cowries live permanently on the corals and graze on mucus, parts of polyps and other tissue. These gastropods are invariably well camouflaged with mantles the color of their hosts and fringes of papillae which resemble expanded polyps.

A large coralliophilid gastropod, *Rapa rapa*, known to collectors as the turnip shell, lives permanently embedded in a calcareous cyst in the base of soft corals. Little is known of its feeding habits but it is undoubtedly a parasite, probably feeding on tissues inside the coral colony.

Also embedded in *Sinularia* are small, highly degenerate parasites which belong to an undescribed family of copepods. A thin-shelled sponge barnacle (*Acasta*) also bores into the soft corals but still feeds on plankton like most of the other barnacles.

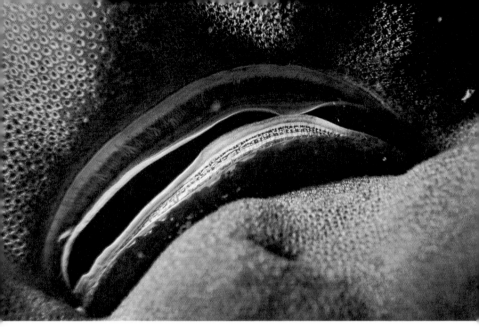

The feeding currents of the scallop *Pedum pedum* modify the growth of the surrounding polyps of its host *Porites,* creating a deep fissure. (4m. Magnetic Island)

Epizoic ark shells *Navicula ventricosa* inhibit the growth of the surrounding polyps, ultimately creating a depression. (4m. Lizard Island)

At least twelve families of Indo-Pacific reef fishes feed on living corals. Parrotfishes (family Scaridae) graze on the corals to extract the endozoic algae and zooxanthellae. Above: *Scarus erythrodon* nibbling at the coral (photo by Allan Power, New Hebrides). Below: Beak-marks of a parrotfish on *Porites*. (Photo by the author, 3m. Palm Island)

Dendronephthya Community

The brilliantly colored spiny soft coral *Dendronephthya*, known as the 'teddy bear' coral, also hosts a range of well camouflaged symbionts.

This coral consists of a long white stalk and vividly colored heads which bear the clusters of polyps. The white stalk is the home of a small white scaleworm (*Paradyte*), a white shrimp (*Periclimenes*) and a white pistol shrimp (*Synalpheus*). The colored area is inhabited by a colored half crab (*Porcellanella*), a majid crab and a brittle star (*Ophiothrix*). The majid crab is beautifully camouflaged with a system of spines which resemble those of the host, exactly matching the host's color. To perfect the disguise this majid even cultures host tissue on its carapace.

SYMBIONTS OF GORGONIANS

The gorgonians, the delicate sea fans and the stouter sea whips and sea fingers, have a horny skeleton composed of a proteinaceous material known as gorgonin. They are occupants of the deeper waters of the coral reef, mainly in areas subjected to moderate currents. The sea fans especially are richly colored in reds and yellows and are one of the most spectacular sights on a coral reef.

They are also very popular hosts for symbionts, partially because their complex body shapes offer many small hiding places and also because they provide firm holdfasts for the filter feeders which exploit strong currents. They also produce copious amounts of mucus and regularly slough their outer cellular layers—both invitations to commensals. The gorgonians are avoided by most predators, possibly because of noxious chemicals they produce. In the Caribbean where they abound they were found in very few of the fish examined. Gorgonian symbionts would therefore be largely immune from direct or indirect predation.

No major studies have been conducted on the symbionts of coral reef gorgonians, but my observations indicate that the community is very rich.

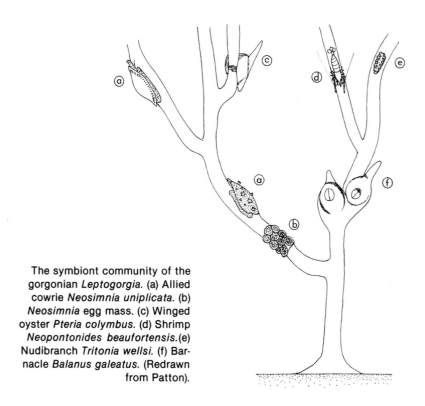

The symbiont community of the gorgonian *Leptogorgia*. (a) Allied cowrie *Neosimnia uniplicata*. (b) *Neosimnia* egg mass. (c) Winged oyster *Pteria colymbus*. (d) Shrimp *Neopontonides beaufortensis.*(e) Nudibranch *Tritonia wellsi*. (f) Barnacle *Balanus galeatus*. (Redrawn from Patton).

Melithaea Community

A Great Barrier Reef sea fan (*Melithaea*) commonly has a winged oyster (*Astropteria*) attached to its base, a halfcrab (*Porcellana*) foraging on its trunk, a true crab *(Xenocarcinus)* clinging to its branches, two small species of shrimps *(Periclimenes* and *Hamodactylus)* on the branchlets and innumerable little brittlestars *(Ophiothela)* coiled around the branchlets. A large spindle egg cowry *(Phenacovolva)* often feeds on the polyps and tissue on the major branches while another smaller species of the same genus grazes on the polyps of the smaller branchlets.

Leptogorgia Community

A detailed study of the symbionts associated with the North American gorgonian *Leptogorgia* illustrates the kinds of relationships gorgonian symbionts have with their hosts. This gorgonian

127

Several different families of fishes live among the branches of corals. Here the hawkfish *Paracirrhites arcatus* sits in a typical pose but will retreat further into the coral if frightened. Photo by Allan Power.

Small gobies of the genera *Gobiodon* and *Paragobiodon* live permanently among the branches of the seriatoporid corals. *Paragobiodon xanthosoma* seen here lives only in *Seriatopora hystrix*. (3m. Lodestone Reef)

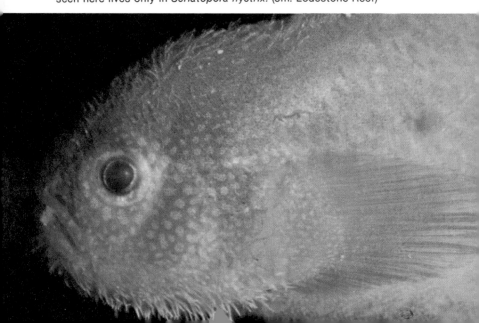

This ophiuroid (unidentified) habitually lives on the bases of the mushroom coral *Herpetoglossa*, possibly feeding on their mucus. (20m. Lodestone Reef)

An isopod *in situ* in a zoanthid.

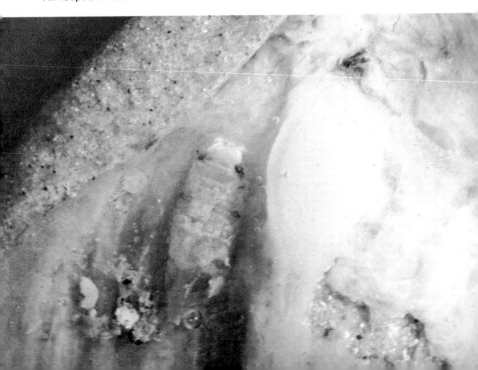

Gorgonians such as this fan coral provide food and shelter for communities of symbionts. (5m. Magnetic Island)

130

hosts two species of copepods, an allied cowry (*Neosimnia*), a nudibranch (*Tritonia*), a shrimp (*Neopontonides*), a barnacle (*Balanus*) and a bivalve (*Pteria*).

The barnacle and bivalve are filterfeeders and do not greatly affect their hosts. However, occasionally the byssus which the bivalve uses to anchor itself may tear away the host tissue (coenenchyme), leaving the horny axis open to fouling by other organisms. The only harm the barnacle does is to induce a minor tissue reaction in the host coenenchyme, which ultimately overgrows it.

By contrast, the other species actually take different materials from the host and may damage it in some way. The shrimp is mainly a scavenger feeding on the host's mucus, adhering organic detritus, food stolen from the polyps and material in the surrounding sediment. The nudibranch actually grazes on host tissue, polyps and coenenchyme, while the cowry tears pieces off polyps and takes mucus and sloughed tissue.

The gorgonian host has a very variable coloration, occurring commonly in shades of yellow and orange and occasionally in red, purple or white. With the exception of the bivalves, all of the symbionts are well camouflaged. The barnacles are hidden by the host tissue which envelops their shells, and they appear only as bumps on the branches. The shrimps have red and yellow chromatophores which they can expand or contract differentially to create the full range of the host's colors with the exception of white. Shrimps generally do not live on white hosts. The nudibranchs do not match their hosts but they instead possess white projections resembling polyps on their dorsal surfaces. They are small enough to be inconspicuous, especially when the polyps are fully expanded. Cowries actually use the host's own pigment, a carotenoid, which is incorporated in their shells. A snail transferred from a host of one color to one of another will end up having a shell of two colors.

IMPORTANCE OF SYMBIOSIS IN THE CORAL REEF ECOSYSTEM

Symbiotic associations are essential for the very existence of corals and may contribute, to a great and unrecognized degree, to the high species diversity of coral reefs.

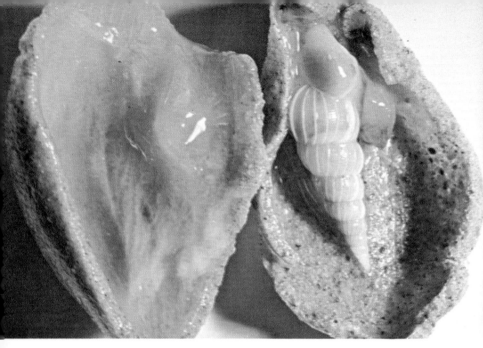

Parasitic *Epitonium in situ* in a zoanthid.

Solitary zoanthids, like other coelenterates, are parasitized by staircase shells, *Epitonium*. (35m. Magnetic Island)

Antipatharian sea whips stretch from the sea floor into the water column. (5m. Broadhurst Reef)

The associations involving the different groups of corals are remarkably similar in detail. The corals attract many other organisms because they provide shelter for the refugees, mucus for the commensals and readily regenerated tissue for the parasites. Since they occupy almost the entire substrate of coral reefs, many different sessile and free-living species have been physically forced into contact with them, and many of these associations have evolved into symbiotic ones.

A study of the ecology and energetics of the coral symbionts may also help us to understand the complex coral reef ecosystem. Coral symbionts, like many other organisms in the coral reef community, have a high species diversity, but with low numbers of most species.

The high diversity of coral symbionts is due to the specialization of their particular ecological niches. Many habitats are provided by, for example, a branching coral. These include the branch tips, outer branches, inner branches, colony base, substrate around the base and water column around the branches. A symbiont may be mobile or sessile; it may live on the surface, in the coenenchyme or in the polyps.

The food available to the coral symbionts is equally diverse. A whole range of plankton and organic materials float in the surrounding water and benthic organisms and detritus lie on or in the surrounding sediment. The coral itself may provide mucus, waste products of digestion and excretion, expelled zooxanthellae and sloughed sheets of epithelial and necrotic tissue for the commensals. Parasites may exploit the coenenchyme, polyps, parts of polyps or body fluids. Thus a single coral may provide niches for scores of other organisms.

How do so many symbionts live in equilibrium with the host? Firstly, many of the symbionts—the refugees and the plankton feeders—do little or no damage to the host. Secondly, the damage inflicted by some of the commensals and all of the parasites must not exceed a rate of regeneration sustainable by the host. This necessitates a strict control of the commensal and parasitic populations.

Populations of all coral symbionts might be controlled at the distributory phase in which their larvae may actually be eaten by

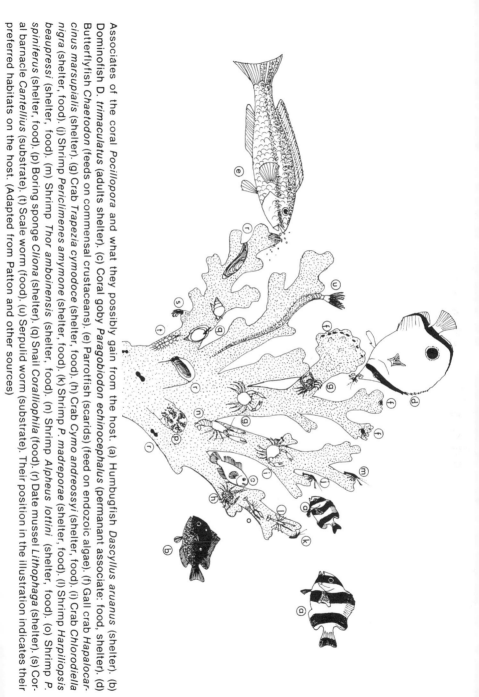

Associates of the coral *Pocillopora* and what they possibly gain from the host. (a) Humbugfish *Dascyllus aruanus* (shelter). (b) Dominofish D. *trimaculatus* (adults shelter). (c) Coral goby *Paragobiodon echinocephalus* (permanant associate: food, shelter). (d) Butterflyfish *Chaetodon* (feeds on commensal crustaceans). (e) Parrotfish (scarids) (feed on endozoic algae). (f) Gall crab *Hapalocarcinus marsupialis* (shelter). (g) Crab *Trapezia cymodoce* (shelter, food). (h) Crab *Cymo andreossyi* (shelter, food). (i) Crab *Chlorodiella nigra* (shelter, food). (j) Shrimp *Periclimenes amymone* (shelter, food). (k) Shrimp *P. madreporae* (shelter, food). (l) Shrimp *Harpiliopsis beaupressi* (shelter, food). (m) Shrimp *Thor amboinensis* (shelter, food). (n) Shrimp *Alpheus lottini* (shelter, food). (o) Shrimp *P. spiniferus* (shelter, food). (p) Boring sponge *Cliona* (shelter). (q) Snail *Coralliophila* (food). (r) Date mussel *Lithophaga* (shelter). (s) Coral barnacle *Cantellius* (substrate). (t) Scale worm (food). (u) Serpulid worm (substrate). Their position in the illustration indicates their preferred habitats on the host. (Adapted from Patton and other sources)

The sea whip shrimp
Pontonides unciger is
well camouflaged on the
sea whip *Cirrhipathes.*
(5m. Broadhurst Reef)

Cottogobius yongei on a sea whip. Sucker-like pelvic fins help maintain the fish in position when water currents are strong. Photo by Walt Deas.

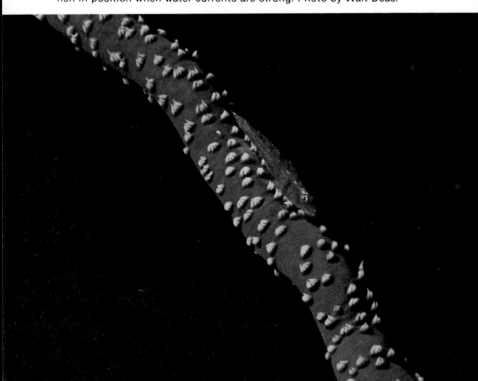

A tree of black coral *Antipathes* is host to a variety of symbionts seeking food and shelter. (15m. Palm Island)

the coral and at the time of settlement when they may be killed or repelled by the coral's defenses. Many of the commensals and parasites also control their own numbers through their strong territorial behavior. Established adults kill or drive newcomers from their territories. Such controls of a species' numbers may also occur in many of the free-living members of the coral reef community.

The coral symbionts show a heavy reliance on coral mucus as an energy source. Coral mucus, rich in wax esters, triglycerides and phospholipids, is now thought to be a basic, easily transferable and recycled energy source for other reef dwellers as well. Mucus has the added advantage in that it can be taken without killing the producer, the hen which lays the golden egg.

Coral symbiology may therefore provide some clues to the puzzling high species diversities, low population densities, and complex energy and resource cycling of the coral reef ecosystem.

To summarize the importance of coral reef symbiosis:

(1) Corals, because of their symbiotic algae, are essentially primary producers.

(2) Corals, because they grow by asexual division, can regenerate much of the damage caused by symbionts and predators.

(3) Coral mucus may be a basic energy source for commensals while on corals and for many free-living species after it is washed off. By thus recycling resources the coral reef ecosystem leaks little energy into the outside waters.

(4) The above properties of corals, together with their ability to produce calcareous skeletons, directly provide many niches for specialized symbionts and indirectly provide niches for all other coral reef inhabitants.

(5) Population regulation by coral symbionts, and perhaps by many free-living inhabitants, ensures that they do not over-exploit resources.

Chapter 5. Soft Bottom Communities

The sandy and muddy sea floors, the benthos, might seem dull and deserted when compared with the vibrant coral reef oasis. They are not, of course—they are merely different. The organisms which live on or in the sediment of the shallow coastal bays, the deeper continental shelf, the continental slopes which plummet into the ocean deeps and the abyssal zone itself are all highly specialized for life on soft bottoms. Many of the occupants of these soft bottoms are mobile, crawling or walking over the bottom, or are fixed, securely anchored by roots. Others live in the sediment or make burrows or tubes in which to shelter.

The soft bottoms, especially those in shallow waters subjected to wave action and strong currents, are not readily exploited by those sedentary animals which require a solid, unyielding surface for attachment. These solid objects present—the skeletons of corals, the shells of molluscs, rocky reefs—are liable to become buried by the mobile sediments and any encrusting organisms may be smothered.

Some sessile species are adapted to take advantage of solid surfaces which appear for a short period and are then covered up again. In North America the Pacific barnacle *Balanus pacificus* is the first colonizer of rocks on the sea floor which have been scoured of sediment. Its larvae must be constantly searching for a suitable place on which to settle for they appear on a new substrate within days and quickly grow and reproduce before it is covered again. This barnacle has been aptly called a fugitive species.

Another alternative for sessile species in this environment is to exploit those species which are specialized for life on soft bottoms. Many algae, sponges, anemones, sea pens and other organisms can

Commensal brittle stars *Astrobrachion* cf. *adhaerens* coiled like snakes around the black coral *Antipathes*. Photo by R.A. Birtles. (16m. Torres Straits)

Epizoic oysters *Dendrostraea folium* cling like little bears to the branches of the valuable black coral *Antipathes*. (Magnetic Island)

The fleshy soft corals host a community of symbionts in and on them. From right to left are *Xenia, Sinularia,* and *Lobophytum.* (Palm Island)

The portunid crab *Caphyra laevis* lives among the long polyps of the soft coral *Xenia elongata.* Note the six 'polyps' on its carapace. (2m. Lodestone Reef)

live in soft sediment because they have elaborate root systems for anchorage. Sessile organisms may also exploit the mobile species, the burrowers and tube makers. We therefore find many epizoic associations in benthic communities.

This trend was vividly demonstrated to me in two unique communities in the tropics of northern Australia, but virtually any mud flat or 'bare' sandy or muddy bottom throughout the world might show similar adaptations in its inhabitants.

MOBILE COMMUNITIES
MOBILE CORAL COMMUNITY
Coral/Worm

Off northern Queensland's deserted coastline lie many high or continental islands surrounded by fringing coral reefs and protected from the Pacific swells by a rampart of long 'ribbon' reefs of the Great Barrier Reef to the east. The continental shelf in this area is shallow, only 30-40m. in depth, and about 50-100km. wide.

On the sea floor surrounding these high islands live some very curious little corals very unlike the reef-builders of the nearby fringing reefs. These corals are solitary (made up of only a single polyp) and have remarkable powers of locomotion.

The mobile coral community was first described by Dr. Tom Goreau of Jamaica and Sir Maurice Yonge of Glasgow, both foremost in their fields of coral biology. These scientists were on board the Belgian Naval frigate *De Moor* during her 1967 expedition to the Great Barrier Reef when she anchored off Lizard Island, a high island named by James Cook in 1770. The rich fringing coral reefs surrounding the group of islands attracted the expedition's attention, and it was not until some days after their arrival that divers decided to investigate the sandy bottom directly under the frigate's keel. There, to their amazement, they found a community dominated by small unattached corals the size of buttons. Even more amazing was the fact that these corals were creeping around the bottom. Several years later I had the opportunity to see this community for myself.

My first sight of the community was memorable. I entered the water about three kilometers from the island and sank through 20 m. of water before the bottom loomed up. Littering the fine

142

sand/mud sea floor were hundreds of small rounded corals, some like mushrooms (*Cycloseris* and *Diaseris*) and others pear-shaped *(Heteropsammia* and *Heterocyathus)*. The former are related to the mushroom or fungiid corals of the reef, but they have spectacular powers of locomotion. They use their tentacles and microscopic whip-like flagellae to creep over the sand and to right themselves if overturned. If the fine sediment buries them they can pump themselves up with sea water and float through to the surface of the sediment.

The pear-shaped *Heteropsammia* has no such inherent powers of locomotion but has achieved the critical mobility needed to exploit this environment by entering into a partnership with another animal. Its partner, the sipunculid worm *Aspidosiphon*, belongs to a group of worms which live in empty snail shells.

The free-swimming planula larva of the coral *Heteropsammia* seeks out a snail shell occupied by a young *Aspidosiphon* and settles on it. The anemone-like polyp begins to lay down a skeleton of calcium carbonate that in time completely covers the shell except for a small hole for the worm, which laboriously drags the coral

The solitary coral *Heteropsammia michelinii*. (A) The coral being dragged along by the sipunculid worm *Aspidosiphon corallicola*. (B) Worm removed from coral. (C) Cross-section of the coral showing the original snail shell embedded in the coral. As the coral grows it enlarges the chamber for the worm. (From various sources)

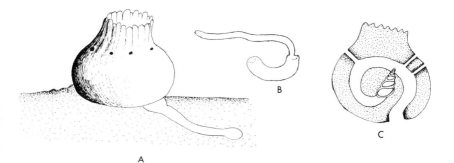

The white stalk of the nephtheid soft coral *Morchellana tixierae* is occupied by various animals. Above it is shown with a half crab (*Porcellana*) and below with one of the pistol shrimps. (15m. Bowling Green Bay)

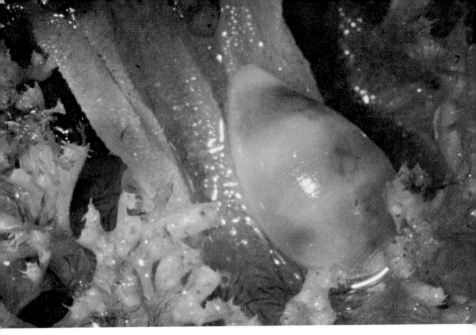

Another symbiont of the nephtheid soft coral *Morchellana tixierae* is this allied cowry. (15m. Bowling Green Bay)

The turnip shell *Rapa rapa* lives enclosed in a chamber at the base of the soft corals. (15m. Coil Reef)

Young *Heteropsammia* polyps early in development. The host worm is visible in one case. (8m. Magnetic Island)

Mature *Heteropsammia*. The coral has grown around the worms shell but leaves a hole at the base for the worm. One coral hosts a *Lithophaga* mussel. (15m. Lizard Island)

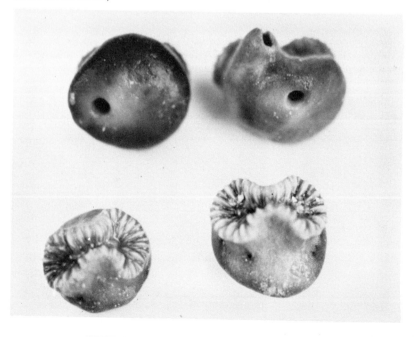

around. The coral can thus be dragged out of smothering sediments and be righted if it is overturned. It also possesses an inherent self-righting capacity as its heavy rounded base, which is occupied by the worm, gives it a low center of gravity and, like the old ballasted children's doll, it tends to fall on its base.

The association is mutualistic: the worm gains a secure, thick-walled home which enlarges as it grows larger. Like the hermit crabs, most other worms must keep changing into larger shells as they grow.

Other mutualistic partners in this complex symbiosis are the zooxanthellae, the plants which live in the coral's tissue and act as powerhouses by trapping sunlight and utilizing the coral's wastes. They probably also aid the coral in secreting its calcium carbonate skeleton.

A small bivalve (*Jousseaumiella*) also lives in cavities in some *Heteropsammia*. It is thin-shelled and probably lives in the coral to gain protection, but may also benefit by feeding on its host's wastes. Small date mussels (*Lithophaga*) bore into the coral's skeleton, and a coral barnacle (*Cantellius*) lives embedded in its surface.

This amazing coral, which is no larger than a thumbnail, is therefore really a very complex combination of organisms, a committee of as many as six different species.

Sponges/Crab

Many of the sponges among the mobile corals were securely rooted in the sediment but other species, not able to grow roots, were also represented; some were even attached to the coral *Heteropsammia*, using it for a hold-fast and for transport.

One day I was observing a coral inching along when a large sponge near me suddenly and mysteriously ran away. I retrieved it and on closer inspection, the reason for its mobility became apparent—a crab was hiding under it, holding onto its base with a pair of legs.

The dromid or sponge crabs carry sponges and occasionally other organisms on their backs for camouflage. Sponges are apparently unpalatable to most predators and the hidden crab is overlooked. Some of the dromids carry whole sponges but others

Soft corals such as this *Dendronephthya* provide a safe refuge and food for many other animals. (20m. Yankee Reef)

A commensal brittle star, an undescribed *Ophiothrix (Acanthophiothrix)*, in *Dendronephthya* host. (20m. Yankee Reef)

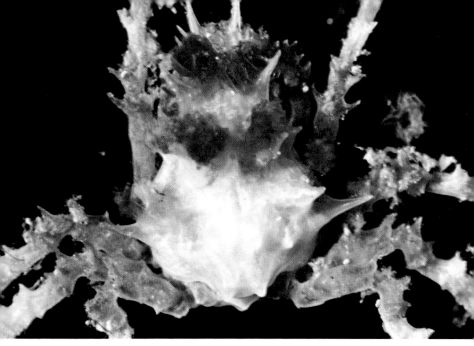

A majid crab removed from its *Dendronephthya* host. The carapace bears spines which resemble the calcareous spicules of the host. (20m. Yankee Reef)

A porcellanid crab *Porcellanella* removed from its host *Dendronephthya*. Its chromatophores have 'painted' a cluster of polyps on its carapace.

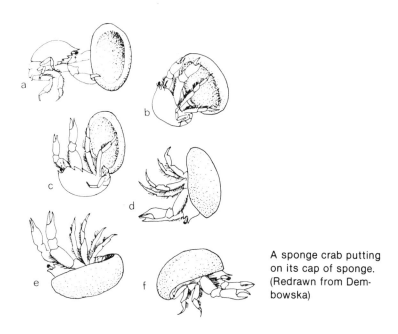

A sponge crab putting on its cap of sponge. (Redrawn from Dembowska)

go to more trouble and cut out a neat-fitting cap which they hold on their carapaces with the specialized upward-pointing legs which are equipped with hooked claws for grasping the sponge.

To put on a new cap a dromid often engages in a series of acrobatics. It will grasp the cap in its claws, roll over on its back, upend itself backwards into the cap and then right itself, finally wearing the cap on its carapace. As with a few other crabs, the eggs of dromids hatch into young crabs, bypassing the larval stages, and these young spend some time sheltering under their mother's abdomen.

The sponges I saw being carried at Lizard Island gain by achieving mobility and a holdfast, but generally the crabs pick or cut sponges which are attached, so the association is usually one-sided.

SPONGES AS HOSTS

The sponges carried by the crabs and those firmly anchored in the mud at Lizard Island were teeming with lodgers, which was expected as sponges throughout the world are well known for their associated communities. They are popular hosts because

150

their complex systems of inhalent and exhalent canals offer excellent protection for animals seeking shelter. The strong currents set up in these canals by the flagellae of the choanocyte cells are exploited by these lodgers for feeding and respiration. The food the sponge traps may be stolen by commensals and parasites, and the large amounts of mucus secreted by the sponge to trap food are also a readily available source of food. The sponges possess few defenses to deter commensals.

The symbionts of sponges include members of most phyla and include bacteria, fungi, algae, anemones, polychaetes and other worms, gastropod and bivalve molluscs, amphipods, isopods, barnacles, crabs and shrimps, ophiuroids, echinoids, holothurians and small fish.

Very large populations may be found in certain sponges—sixteen thousand pistol shrimps (*Alpheus*) have been recorded from a single sponge. I have collected sponges which on close examination were little more than barnacles (*Acasta*) packed shell-to-shell

The pistol shrimp *Synalpheus neomeris* is another animal that lives in the hollow stalk of *Solenocaulon*.

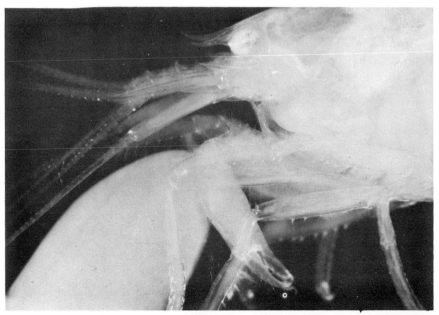

The allied cowry *Pseudosimnia* on the stalk of a *Dendronephthya*. Photo by
R.A. Birtles. (12m. Torres Straits)

Gorgonian sea fans and sea whips act as hold-fasts for some plankton
feeders. (20m. Lizard Island)

The sea finger, the gorgonian *Rumphella.* (5m. Magnetic Island)

The gorgonian sea fan *Rumphella aggregata* harbors various types of animals. Note the banded echinoderm in the center of the photo. (8m. Broadhurst Reef)

and covered with a thin veneer of host tissue. Likewise I have collected sponges and placed them in aquaria overnight and next morning found many hundreds of ophiuroids which had left the canals of their already-putrifying hosts. Some sponges host large populations of a single species whereas others host a mixed community.

Sponge Commensals

Those filter-feeders which obtain shelter and feeding currents include the sponge barnacles (*Acasta*) and bivalves of the family Vusellidae, known as sponge fingers. A sponge may pump ten thousand times its own volume of water each hour, and it is a simple matter for filter feeders to tap these currents. The sponge pumps in water and sieves out plankton and detritus through small pores, the material sticking onto sheets of mucus. Crabs (brachyurans), half crabs (porcellanids), mantid, alpheid and palaemonid shrimps, polychaetes, ophiuroids and holothurians may all feed on the trapped food and the mucus sheets.

No energetics studies have been conducted on these associations, but to support so many commensals sponges must be very efficient feeders with an unexpectedly high metabolism.

Sponge Mutualism

Sponges benefit from a relationship with symbiotic dinoflagellates, the zooxanthellae. These algae take wastes (carbon dioxide and metabolic products) from their hosts and release oxygen which is, in turn, utilized by the sponge.

Sponge Parasites

Undoubtedly many of the 'commensals' have a parasitic association, but the sponges have more readily identifiable parasites such as bacteria, fungi and nudibranchs and other gastropods.

The fungi may be very destructive. An epidemic in the late 1930's destroyed almost all of the stocks which were the basis of the rich and colorful bath-sponge fishery of Florida. It has taken many years for them to recover.

The half crab *Porcellanella trilobata* lives on the major branches of *Solenocaulon.*

This larger half crab *Aliaporcellana telestophila* lives in the hollow stem of *Solenocaulon.*

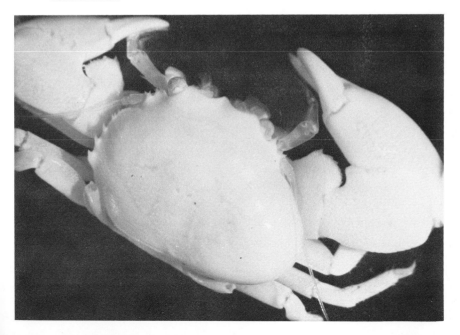

Ophiothela cf. *hadra*, a commensal brittle star, wraps its long arms around the branch of a *Rumphella*. It probably steals its host's mucus.

Winged oysters *Astropteria lata* use *Rumphella* and other gorgonians as holdfasts.

The transparent gorgonian shrimp *Periclimenes psmathe* on the sea fan *Melithaea ochracea*. (5m. Magnetic Island)

P. psmathe has red and yellow chromatophores to match the color of its host. The eye stalk, highly magnified.

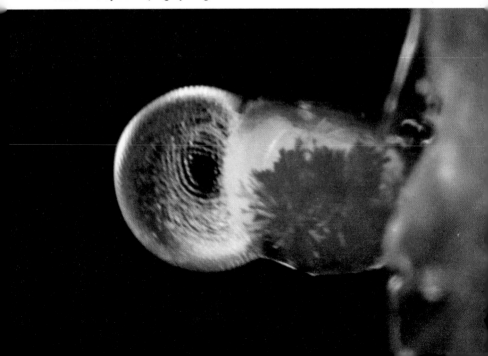

HYDROID COMMUNITY

Clumps of feather-like hydroids grow on dead shells and corals in the mobile coral community of Lizard Island and in many other communities. They, like the sponges, host a variety of symbionts.

A variety of filter feeders use the Lizard Island hydroids for attachment. Small well-camouflaged wing shells (*Electroma*) and grotesque skeleton shrimps (*Caprella*) are almost indistinguishable among the branches. White sea anemones, white half-crabs and orange ophiuroids were more conspicuous symbionts.

Most of the hydroids' symbionts use them for shelter because the individuals of the hydroid colonies, the hydranths, have batteries of defensive and offensive stinging cells, nematoblasts. These cells do not deter parasitic or predatory aeolid nudibranchs which eat the hydranths and use their nematoblasts for their own defense.

BURROW DWELLERS

Soft bottoms from the intertidal to the abyssal zone are riddled with the burrows of many different organisms. Pistol shrimps (alpheids), mantis shrimps (stomatopods), polychaete and echiuroid worms, holothurians, echinoids and many others excavate burrows which are used in turn by other organisms.

Fish/Shrimp Partnerships

Gobies of the genera *Cryptocentrus, Vanderhorstia* and *Psilogobius* habitually live in the burrows of the pistol shrimps *Alpheus* and *Synalpheus*. The burrow is used as a temporary refuge during the day and as a permanent resting place at night. The association is mutually beneficial as the shrimp, which has poor vision, depends on the alert fish for a warning of the approach of predators.

While the shrimp is busily engaged in excavating its burrow and feeding, the goby lies perched at the entrance. When danger threatens it warns its partner by giving a barely perceptible flick of its tail which the blind shrimp senses with its antennae which are in constant contact with the fish. The shrimp then darts into the hole and the fish follows. When the danger passes the fish and then the shrimp reappear.

The symbiosis is obligatory for both partners since adults of those species are never found apart. It begins at an early stage when very young fish pair up with very young shrimps.

Like the gregarious free-living burrowing shrimps, the goby-shrimps often live in communities, but their burrows are generally more widely spaced. The difference in spacing is due to the gobies' territoriality. However, if these gobies' territories are compared with those of free-living goby species it will be found that the former are much smaller. This difference has been attributed to the fact that the symbiotic gobies spend much of their time in their associates' burrows and therefore do not engage in as much boundary fighting as the free-living gobies. The overall dispersion of the goby/shrimp burrows is therefore an interesting compromise between the spatial preferences of each group.

Burrow-Dwelling Bivalves

The members of the bivalve family Lasaeidae frequently live in burrows excavated by other organisms. The small, almost shell-less *Phlyctaenachlamys lysiosquilla* lives in the burrows of the large mantis shrimp *Lysiosquilla; Ephippodonta* lives in shrimp burrows; *Lepton* and *Mysella* live in sipunculid worm burrows. Others of this family have even more intimate associations with their hosts. *Jousseaumiella* lives inside the coral *Heteropsammia* and *Entovalva* lives inside holothurians as a parasite.

SOLENOCAULON COMMUNITY

A second community even more vividly illustrates the high incidence of epizoism and other partnerships in soft bottom communities.

This community was discovered by a colleague and me in a mangrove-fringed muddy bay in northern Queensland. This bay, named Bowling Green Bay by the explorer James Cook, is shallow and extremely muddy due to several streams emptying into it. Underwater visibility is rarely more than one or two meters, making visual observations very difficult.

The beautiful shrimp *Hamodactylus boschmai* lives on the thicker branches of *Melithaea*. (5m. Magnetic Island)

Detail of the chromatophores on the cephalothorax of *H. boschmai*.

Gorgonian shrimp *Dasycaris ceratops*. (30m. Lodestone Reef)

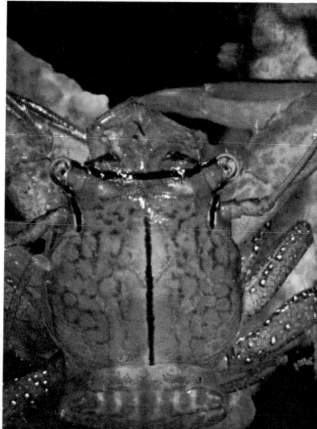

Porcellanid crab
Porcellana furcillata
lives at the base of
Melithaea. (5m.
Magnetic Island)

Our discovery occurred after a shrimp trawler brought up a portion of a white organism which could not be identified. We later took a motor boat to the remote bay and began a series of underwater searches, although hampered by the poor visibility. Divers were towed on a long rope behind the boat and several kilometers of sea floor were searched unsuccessfully.

The sea floor was remarkably uniform and barren. Only one crinoid and a sea whip attached to a dead bivalve were seen during hours of searches, but finally a diver signaled success and we all entered the water. Below us was a forest of meter-high stalks, pure white in color. They were obviously a coelenterate, but none of us could even guess what kind. They happened to be a new species, now called *Solenocaulon grandis*, an aberrant gorgonian.

The gorgonian comprised a central rod, hollow above and solid below, where it became flat for anchorage in the mud. Large white polyps rose from the stalk and its alternate side branches. Remarkable though this creature was, even more remarkable were the teeming symbionts which literally covered it.

On each colony hundreds of small scale worms (*Paradyte*) and shrimps (*Periclimenes*) hid among the polyps, dozens of ophiuroids

The small shrimp *Periclimenes* (undescribed species) lives on the surface of *Solenocaulon*.

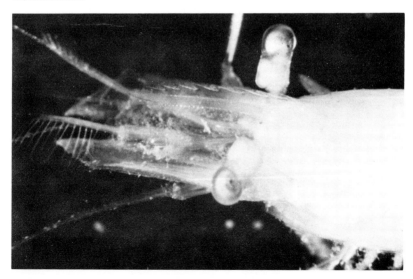

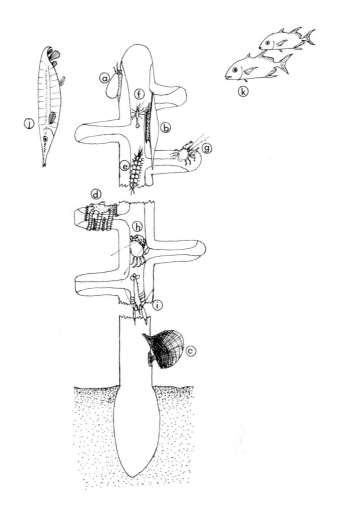

The symbiont community of the gorgonian *Solenocaulon grandis.* (a) Allied cowrie *Primovula rutherfordiana.* (b) Spindle cowrie *Phenacovolva rosea.* (c) Winged oyster *Astropteria lata.* (d) Ophiuroid *Ophiosammium semperi.* (e) Polynoid *Paradyte.* (f) Shrimp *Periclimenes* sp. (g) Half-crab *Porcellanella trilobata.* (h) Half-crab *Aliaporcellana telestophila.* (i) Pistol shrimp *Synalpheus neomeris.* (j) Razorfish *Aeoliscus.* (k) Juvenile trevally (carangid)

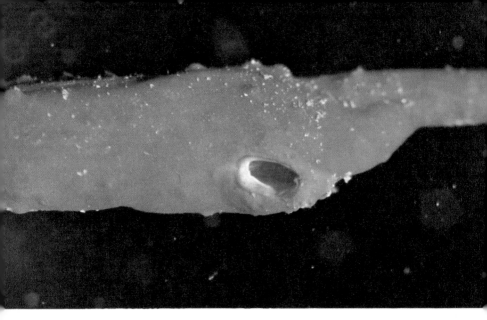

Balanid barnacle overgrown by its gorgonian host. (5m. Magnetic Island)

Commensal brittlestars *Ophiothela* on *Melithaea* (5m. Magnetic Island)

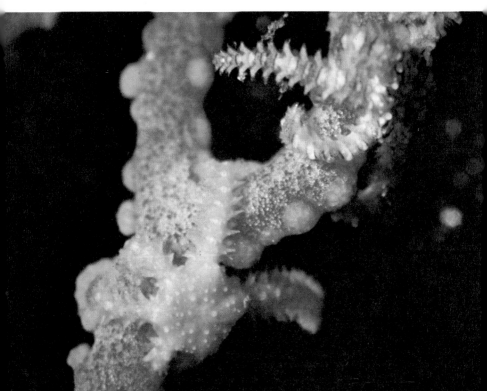

The spindle cowry *Phenacovolva rosea* is elongate for life on the narrow branches of the gorgonians. (40m. off Magnetic Island)

The spindle cowry *P. angasi,* here greatly enlarged, lives on the narrower branches of *Melithaea*. The red, white and yellow mantle camouflages it on its host. (5m. Magnetic Island).

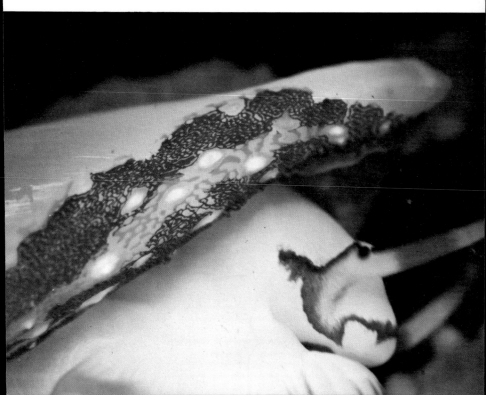

(*Ophiopsammium*) were wrapped around the branches and hundreds of tiny amphipods and copepods swarmed over it. Inside the hollow section were many small half-crabs (*Porcellanella, Aliaporcellana*) and many pistol shrimps (*Synalpheus*). At the base was a winged oyster (*Astropteria*). Elongate spindle cowries (*Phenacovolva*) grazed on the tissue of the central stalk and small allied cowries (*Primovula*) grazed in the more inaccessible places. All of these animals were the pure white color of the host.

A flattened razorfish (*Aeoliscus*) hung head down next to the stalk of one of the colonies, trying to escape detection. Small schools of juvenile trevallies circled, also seeking shelter.

How can so many symbionts exploit a single host without damaging it? The answer is probably the same as that given in the previous chapter: fine niche division and population regulation.

The symbionts match the color of their host in different ways. Some, for example the cowries, take pigment from the host and incorporate it into their shells and tissues. I had seen the same spindle cowry, *Phenacovolva rosea*, on red gorgonians before and it had matched its hosts exactly. Some of the crustaceans expand and contract color cells (chromatophores) to match their host or are transparent.

The *Solenocaulon* community is one of the most unusual I have encountered. It is virtually the only sessile organism able to colonize large areas of the silty bay and because of this is a haven for all those organisms seeking shelter. It is like an oasis of palms in a featureless desert or a coral atoll in a relatively barren tropical ocean.

Chapter 6. The Open Sea

If solid surfaces for settlement or shelter are uncommon on the soft bottoms, then they are very much rarer in the water body above.

A drifting coconut, a piece of almost submerged volcanic pumice stone, a detached clump of the seaweed *Sargassum*, a marker buoy, or even a discarded glass or plastic bottle—all are rapidly colonized by the microscopic spores of algae and the larvae of sessile hydrozoans, bryozoans, bivalves and barnacles. Fish, from fry to *Rhincodon*, the giant whale shark, hover around drifting objects seeking shelter or food. Commercial and sport fishermen are well aware that an almost inconspicuous floating object will often attract large pelagic fish.

The famous *Kon-Tiki* raft became a haven for the homeless refugees of the open sea during its epic voyage halfway across the Pacific in 1947. The big balsa logs accumulated algae, small crabs and barnacles, and many fish gathered under it or followed it at a distance.

Thor Heyerdahl, the expedition leader, was acutely aware of the vastness of the ocean and the precarious existence of those in it and on it, including themselves. Epitomizing this insecurity was a crab which they came across clinging tenaciously to a small floating feather molted from some sea bird many hundreds of kilometers from any land. As the lumbering *Kon-Tiki* drifted by the little crab deserted its sinking feather for the greater security offered by the balsa logs.

To appreciate the great pressures acting on organisms seeking shelter in the vast open sea, one should drift as the *Kon-Tiki* did or spend long hours peering over the side of a becalmed or slowly drifting yacht. As Heyerdahl said:

A forest of gorgonian sea whips *(Junceela)*. (15m. Lizard Island)

Crinoids frequently use sea fans for holdfasts in areas of strong currents. (10m. Brewer Reef)

A community of mobile corals on a mud-sand bottom. The mushroom coral *Cycloseris cycloides* is naturally mobile but the pear-shaped *Heteropsammia michelinii* relies on its partner, the worm *Aspidosiphon corallicola*, for transport. (15m. Lizard Island)

"The sea contains many surprises for him who has his floor on a level with the surface, and drifts slowly and noiselessly ... We usually plough across it with roaring engines and piston strokes, with the water foaming around our bows. Then we come back and say that there is nothing to see in the ocean."

Thus during the 1960's Hawaiian based biologists Gooding and Magnuson conducted a series of drifts on a raft designed with an underwater observation chamber so that they might investigate the ecological significance of drifting objects to pelagic fish. Over thirty species of fish associated with the raft during its drifts in the open sea near Hawaii and in the Central Pacific. Some of the fish (flyingfish, marlin and tuna) were transients which investigated the raft and left. Others (manta rays, whale sharks, blue sharks, dolphins and pilot whales) were more regular visitors, but many (wahoo, trevallies, pilotfish, dolphin, amberjacks, white-tip sharks and driftfish) were permanent or semi-permanent residents.

The fish, reptiles and mammals which associated with the raft gained protection from predation and benefitted from a concentrated food supply. Some might have used the raft as a cleaning station or to scratch themselves against to remove ectoparasites. Gooding and Magnuson concluded:

"A floating object in the pelagic environment provides a relatively rare *superstrate* in an environment notable for its homogeneity. This superstrate has some of the same ecological significance to certain pelagic fishes that a substrate has to inshore fishes."

Not only do many open-water animals associate with such superstrates, but they also associate with larger free-swimming animals: animate superstrates. A living superstrate has certain advantages over a non-living one. A strongly swimming host provides protection from predators and also feeding and respiratory currents. It also actively maintains its position in a given environment, whereas non-living superstrates ultimately sink or become stranded on a shore. Large, dangerous, toxic or poisonous pelagic animals are avoided by predators, and many smaller animals sheltering in or near them benefit. Sessile animals are also given a surface for settlement and gain mobility and a mechanism for dispersal of their young.

Thus a whole range of diverse partnerships has evolved among the organisms of the open waters of all of the world's seas. These organisms include the surface dwellers (pleustron), the drifting organisms (plankton), and the actively swimming organisms (nekton).

JELLYFISH/FISH ASSOCIATIONS

The jellyfishes, siphonophores (such as the Portuguese man-o-war *Physalia*) and scyphozoans (the jelly blubbers) attract a number of small pelagic fish. More than fifty species of fishes belonging to nine families have independently established partnerships with several dozen species of jellyfishes.

The symbioses are primarily for shelter. Jellyfishes are armed with batteries of stinging nettle cells, the nematoblasts, which are used in the capture of their prey and for defense. Certain small fishes, either the pelagic young of larger species or the adults of small species, accompany the drifting or slowly swimming jellyfishes, at times venturing so near that they actually make contact with the dangerous tentacles. Although the small fishes which are the normal prey of the jellyfishes are quickly killed by the venom of the nematoblasts, the consort fishes are unharmed.

The nematoblasts of the jellyfish, when triggered, explode, firing a venom- filled harpoon.

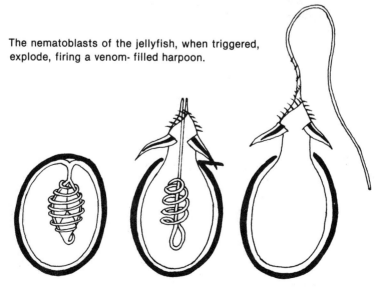

The *Heteropsammia* may have evolved in this way. A dead snail's shell serves as a home for a sipunculid worm and as a holdfast for a mussel and hydroids (stained red). The ancestral coral may have similarly settled on such a shell. (8m. Magnetic Island)

A sponge, unable to find any other solid object, uses a *Heteropsammia* as a holdfast. Surfaces for settlement are often smothered by the shifting sediment. (15m. Lizard Island)

An undescribed association between a soft coral and sponge. The two species are very frequently found together off temperate Australia. Photo by T. Done. (10m. Solitary Island, New South Wales)

The sponge crab *Dromidiopsis edwardsi* cuts a snugly fitting cap of sponge and holds it on its carapace. (30m. Magnetic Island)

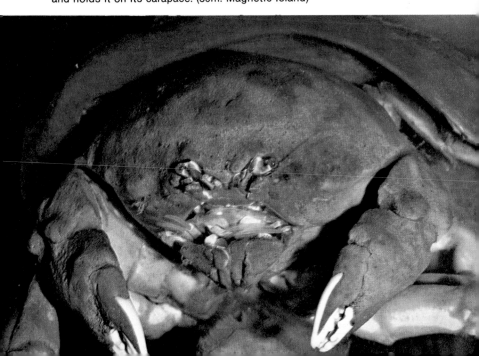

TEMPORARY CONSORTS

In the majority of cases the jellyfish/fish symbiosis is temporary—only the vulnerable pelagic juveniles seek the protection of the jellyfish while the adults rely on other defenses. The adults of these consorts may be either pelagic (trevallies or carangids), demersal (cod or gadids), or abyssal (squaretails or tetragonurids).

The symbiosis, even though it is only temporary, may be complex and dynamic. The North American harvestfish (*Peprilus*) at first seeks only the protective mantle afforded by the tentacles, but as it grows it begins to eat parts of its host and finally consumes it. The adult fish later becomes a predator of jellyfishes. Thus the partnership changes, starting with inquilinism, then to commensalism or parasitism and then to predation.

Clouds of small fishes may shelter under large hosts. Over three hundred juvenile trevallies have been sighted sheltering under a single jellyfish. Often the fishes are not specific in their host preferences, but in some cases they are and the biologies of the two are closely linked.

Nearly every wet or monsoon season in northern Queensland large aggregates of lion's mane jellyfish (*Cyanea capillata*) appear in the muddy coastal waters, probably feeding on plankton and small fishes thriving in the nutrient-laden waters. Perhaps coincidentally, perhaps planned, a number of trevallies breed some time before and their juveniles shelter among the fine tentacles of the jellyfish. At times I have seen juveniles of five species of trevally, a pomfret and a coral reef damselfish all sheltering under a single large *Cyanea*—some thirty or forty fish in all. Once, when I netted a host and its consorts the fish unavoidably came in contact with the fine, long tentacles. However, only the damsel (*Chromis*) and the pomfret were seriously stung; both died shortly after while the trevally species were unaffected.

PERMANENT CONSORTS

Some fishes, mainly some of the butterfishes (Stromateidae), have become permanent associates of jellyfishes. Best known of these is the Portuguese man-o-war fish (*Nomeus*), a beautiful little silver and indigo long-finned fish which lives among the tentacles of its host, *Physalia*, and sometimes with the scyphozoan jelly

174

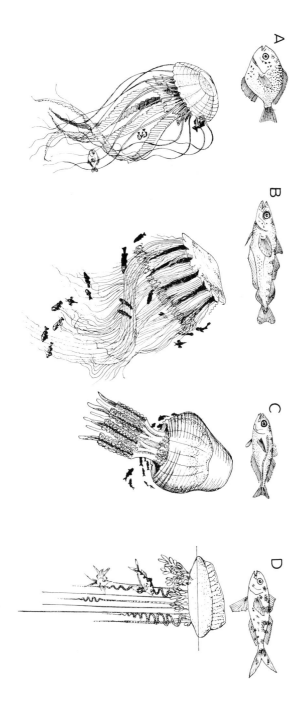

Associations between fish and jellyfish. (A) Harvestfish *Peprilus alepidotus* and sea nettle jellyfish *Chrysaora quinquecirrha*. (B) European whiting *Gadus merlangus* and pink jellyfish *Cyanea capillata*. (C) Mediterranean scad *Trachurus mediterraneus* and *Rhizostoma pulma*. (D) Portuguese man-o-war fish *Nomeus gronovii* and *Physalia*. (Redrawn from Mansueti and other sources)

175

Dromidiopsis dorma feeds quietly, secure in the protection afforded by its sponge cap. Photo by Scott Johnson. (Makua, Oahu, Hawaiian Islands)

A sponge crab carries a sponge, itself the host of a variety of symbionts. (20m. Lizard Island)

Above and Below: Commensal crabs live in the sponges which are in turn carried by the sponge crabs.

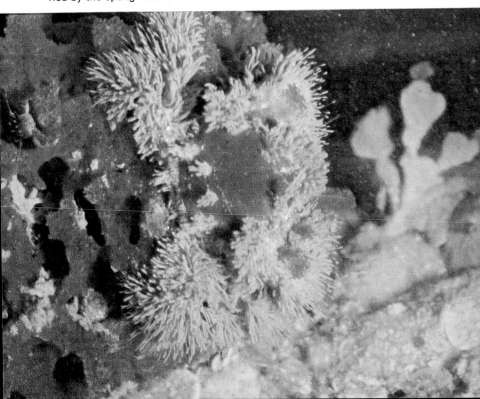

blubbers. *Nomeus* eats planktonic material and occasionally indulges in parts of its host, even the tentacles, without being severely stung.

Other butterfish, such as the medusaefish (*Psenes*), are obligatory symbionts of the scyphozoan jelly blubbers and many examples may be found in the world's seas. Little is known of these paradoxical fishes, either their feeding habits, their relationships to their hosts or their reproductive cycles.

EVOLUTION OF CONSORTFISH

The evolution of the jellyfish/fish partnerships can be traced through a spectrum of associations which grades from casual or opportunistic ones to highly refined obligatory ones.

The former associations are but one step removed from the fish/flotsam associations. Often the young of a species associate with a jellyfish only if they chance upon one, but at other times

The raftfish *Mupus,* a consort of scyphozoans and siphonophores. (Evans Head, New South Wales)

they might shelter near drifting flotsam or travel in tight schools. Although such associations may not be critical for the survival of the species, they probably enhance it.

The association is more refined in fishes such as the harvestfish as the young totally depend on the jellyfish for protection and food. However, after a time the symbiosis breaks down for the fish ends up by eating its host.

The man-o-war consort (*Nomeus*) and the medusaefish (*Psenes*) are the most specialized of all for their association extends past the juvenile stage to become a lifelong one. The adults cannot live apart from their hosts but the jellyfish can survive without their consorts—the symbiosis is one-sided. But does the host gain anything? Perhaps it benefits in small ways, for example by having some of its parasites removed. Popular belief has it that the consort fishes have a reciprocal relationship with the jellyfish for they are thought to act as decoys, luring other fishes to their deaths among the tentacles. In return they are believed to receive part of the prey as reward. This has yet to be substantiated, but like many other popular beliefs may not be without some substance.

IMMUNITY TO STINGS

The stinging nematoblasts of coelenterates are venom-filled cells containing invaginated harpoons and tubes. The cells are discharged by hair-like triggers on their surfaces. If an organism brushes a hair it stimulates the cell to suddenly contract, thus pushing out the sharp harpoon which may lodge in the prey. Venom is injected via a tube through the hollow harpoon.

How do consort fishes avoid being stung to death by their hosts? Probably some of the less specialized fishes simply avoid brushing the tentacles. The little coral reef damselfish and the pomfret which were inadvertently forced onto the tentacles of the *Cyanea* could not have possessed the protective factor which the juvenile trevallies had.

Fishes such as *Nomeus* and *Peprilus* which feed on their hosts' tentacles must also possess some kind of immunity, not only on their body surfaces but also in their mouths and stomachs. Digestive enzymes in the stomach would break down the venom, ultimately rendering it harmless.

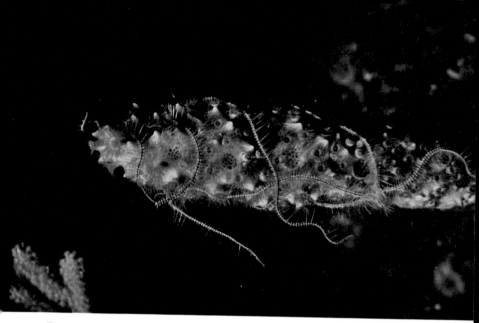

The sponge *Gelloides ramosa* hosts an array of animals. Seen here are brittle-stars clinging to it (at night) and zoanthids *Parazoanthus parasiticus* attached to it. Photo by Dr. Patrick L. Colin. (15m. Crooked Island, Bahamas)

Commensal brittlestars *Ophiothrix ciliaris* on a sponge. Photo by R.A. Birtles. (4m. Torres Straits)

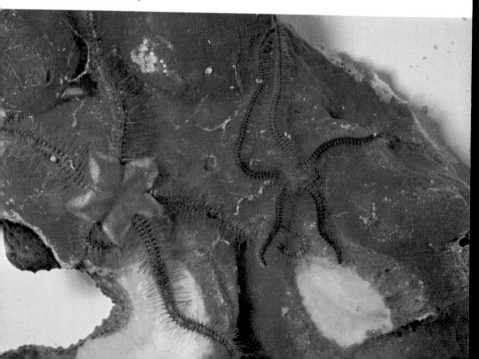

The small white objects in the atria of this sponge (*Neofibularia nolitangere*) are parasitic polychaete worms, *Syllis spongicola*. This worm infests other sponge species as well. Photo by Dr. Patrick L. Colin. (15m. Discovery Bay, Jamaica)

An association between the boring sponge *Cliona delitrix* and the zoanthid *Parazoanthus parasiticus*. The substrate is usually one of the massive corals. Photo by Dr. Patrick L. Colin. (15m. Discovery Bay, Jamaica)

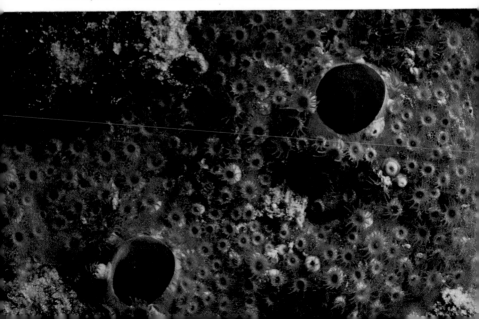

Microscopical studies of unspecialized fishes forced against the tentacles of jellyfish show that the venom-filled nematoblasts are discharged and penetrate their skins, whereas studies of specialized fish similarly treated show that the nematoblasts are not discharged at all. It is therefore thought that specialized fishes have a protective coating which prevents the triggering of their hosts' nematoblasts. This coating is probably the host's own mucus which the consort may steal for a biochemical disguise. The jellyfish probably "recognizes" itself by such coatings, otherwise it would fire nematoblasts at itself every time its tentacles brushed together. The process is probably similar to that of the colorful Indo-Pacific clownfish, associates of the giant anemones.

Might it be possible that we could use a similar material to prevent jellyfish stings? Jellyfish are a danger wherever man works or bathes in the sea, but nowhere is the danger more acute than in tropical Australia and the islands to its north.

The box jellyfish (*Chironex fleckerii*), probably the most venomous of all animals, has been responsible for many fatalities in these waters. Stings from their long tentacles have been likened to burns from red hot pokers and victims have died from shock within one or two minutes. Coastal bathing during the long, hot summer months is not recommended and the subsequent loss to the tourist industry—the coast of Queensland serves the Great Barrier Reef trade—is reckoned to be millions of dollars per annum.

OTHER ASSOCIATES OF JELLYFISHES

Jellyfishes also attract a range of other organisms from refugees like the consortfishes to parasites and predators.

Amphipods

Small amphipods also shelter under the bells of jellyfishes and live in their subgenital pits. *Hyperia galba*, notable for its large and beautiful green eyes, lives under *Cyanea* and *Rhizostoma*. Gut contents of these amphipods contain nematoblasts, mucus and scraps of the host's food, suggesting that they are parasites or commensals.

182

Consortfishes sometimes feed on these crustaceans, thereby performing a valuable cleaning function for their hosts. *Hyperia* are known to be an important food source for young cod (*Gadus*), consorts of Irish Sea jellyfishes.

Shrimps

Many small shrimps live among the tentacles of sea anemones, and some of these species are occasionally found on the tentacles of the jellyfishes. It seems that for shrimps like *Periclimenes holthuisi* a jellyfish is a second choice as a host if it cannot find a suitable sea anemone. The jellyfish host may be important in the shrimp's dispersal for their primary hosts, the sea anemones, move little during their adult lifetimes.

Many small shrimps can be found among the tentacles of sea anemones and jellyfishes. The sea anemones seem to be preferred over the jellyfishes. Here *Periclimenes yucatanicus* (two) and *Thor amboinensis* share an anemone in an aquarium—at least temporarily. Photo by R.P.L. Straughan.

A dense population of sponge barnacles, *Acasta*. One has been dissected out of the sponge. (15m. off Magnetic Island)

A sponge can contain dozens of species and thousands of individuals. This sponge contained several species of crabs and half crabs, shrimps and pistol shrimps, bivalves, barnacles and a variety of worms. (30m. Lodestone Reef)

The shrimp *Rhyn-chocinetes rigens* on the sponge *Agelus* sp. Night photo by Dr. Patrick L. Colin. (22m. Spring Gardens, Jamaica)

Some crabs may live on sponges as well as carry sponges on their backs. This small decorator crab (family Majidae) bears sponges on its carapace but is not well covered by them. Photo by Dr. Patrick L. Colin. (15m. Mona Island, Puerto Rico).

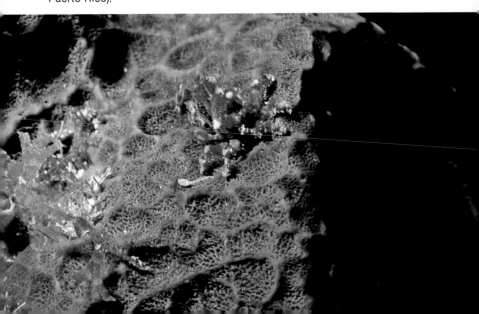

Spiny Lobsters

Some of the delicate, transparent, leaf-like larvae of the spiny lobsters are parasites or parasitoids of jellyfishes. The phyllosomas, as these larvae are called, are planktonic but have been found attached to jellyfishes. They have voracious appetites—or perhaps jellyfishes have little nutritive value—as a small phyllosoma may ultimately consume a jellyfish many times its size.

Barnacles

Another unlikely associate of jellyfishes is the goose barnacle *Alepas*. The transparent, shell-less barnacle settles on the bells of large jellyfishes such as *Cyanea*, thereby gaining protection from predators, a mobile superstrate and feeding and respiratory currents. It has the well-developed plankton-feeding appendages (cirri) characteristic of barnacles, but it may also feed on its host. Some of the less specialized goose barnacles (*Lepas*) are known to tear off pieces of jellyfishes. *Alepas* is uncommon and its life cycle must be closely linked to that of its host.

Gastropods

A small pelagic nudibranch, *Phylliroe*, has a similar relationship to medusae as the spiny lobster phyllosoma larvae have. For many years it was presumed that the jellyfish was a parasite of the nudibranch, but the converse is the case: the nudibranch larva attaches to the jellyfish and feeds parasitically. When it is large enough to swim it consumes its host and becomes free-living.

The beautiful blue and silver nudibranch *Glaucus*, named after the ancient Greek sea god, has a curious association with jellyfishes which might be described as symbiotic. *Glaucus* eats coelenterates such as the Portuguese man-o-war *(Physalia)*, the by-the-wind sailor *(Velella)* and the round-rafted *Porpita*, and, like other aeolid nudibranches, somehow prevents its prey's venomous nematoblasts from discharging.

The prey is digested but the undischarged nematoblasts strangely remain untouched. These cells are then transported from the gut into the frilled appendages called cerata where they once again assume a defensive role. Like certain other aeolid nudibranchs, *Glaucus* may actually sting a human if it is picked up.

Young of another mollusc, the octopus *Tremoctopus violaceus*, similarly use Portuguese man-o-war nematoblasts for defense. This octopus collects portions of tentacles and holds them in its arms for use as weapons of offense or defense.

BARREL SHRIMPS AND SALPS

The transparent salps are often mistaken for jellyfishes but are actually closer relatives of the vertebrates. A small amphipod (*Phronima*) is a parasitoid of the luminous salp (*Pyrosoma*). Young amphipods enter the salp and begin feeding on its internal organs; by the time they reach maturity they have consumed all the host's soft parts and only the transparent barrel-like covering remains. *Phronima* then rears its young in the protective barrel. The amphipod gains food, shelter for itself and its young and a mechanism of dispersal for its species.

In another raft drift, this one the antithesis of Heyerdahl's *Kon-Tiki* expedition, Jacques Piccard spent many hours watching bar-

Barrel-shrimp. The amphipod *Phronima* hollows out the body of the salp *Pyrosoma* and rears its young in it. (Redrawn from Gotto)

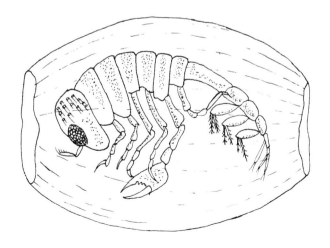

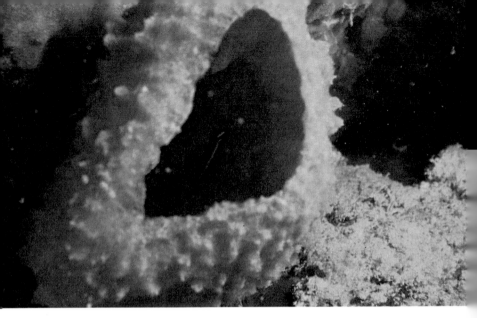

Some gobies of the genus *Gobiosoma* are considered sponge dwellers. Here *Gobiosoma xanthiprora* can be seen in the sponge *Callyspongia plicifera*. Photo by Dr. Patrick L. Colin. (Dry Tortugas, Florida)

An occasional visitor on sponges is *G. atronasum*. Just in front of it is another goby *Coryphopterus personatus*. Photo by Dr. Patrick L. Colin. (Exuma Sound, Bahamas)

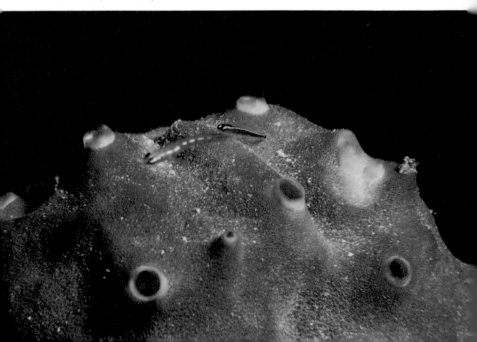

Sea anemones using a colonial hydroid for settlement. Photo by L. Zell. (15m. Palm Island)

rel amphipods from the comfort of his bunk. His 'raft' was the space-age mesoscaph *Ben Franklin*, a submarine designed to drift with the currents at a depth of 200-500 m.

On the day that man first walked upon the moon Piccard spent many hours marvelling at the amphipod inside its transparent home. His thoughts mirrored those of Heyerdahl as he wrote of *Phronima*:

"What mysteries and what grandeur, just as there are mysteries everywhere else to be sure. Here we have time to look, to observe, to think."

SYMBIOTIC BARNACLES OF THE OPEN SEA

Their sessile ways of life would apparently prohibit adult barnacles from exploiting the plankton-rich surface waters of the open sea, but this is far from the truth. The goose or pedunculate barnacles, a major group, have evolved primarily as occupants of the surface waters for they are specialized for settling on drifting objects. As a natural progression of this many goose barnacles and some acorn barnacles are specialized to settle on animate superstrates—the jellyfishes, fishes, reptiles and mammals which live in the surface waters of the open sea.

Much of our knowledge of the symbiotic barnacles comes from the great Charles Darwin, who had a life-long love of the group. Darwin's interest in the group was sparked when, during his epoch-shattering voyage on H.M.S. *Beagle*, he discovered a small burrowing barnacle which he could not place in any known order. Thus began a full time study of the biology of the Cirripedia which lasted from 1846-1854. His tomes remain the definitive works on the subject to this day.

An interesting story is told which demonstrates Darwin's preoccupation with barnacles. One of his sons was visiting the home of a friend and, after being shown his young acquaintance's home, asked: "Where does your father keep *his* barnacles?"

Darwin's study of the barnacles was invaluable for he came to understand and formulate the concepts of species and varieties and variation within species. Here he found evidence to support his unpublished ideas on evolution. Variation, displayed in his barnacles, could be the vital raw material of evolution!

190

Nearly 43% of all acorn barnacles described by Darwin lived as symbionts on other organisms. About one quarter of these lived on open water vertebrates.

Barnacles of Marine Reptiles and Mammals

By entering into symbiotic associations with mobile animals the sessile barnacles have themselves acquired mobility without the expenditure of energy. They can feed in the rich surface waters of the open sea, disperse their young and gain protection from their greatest enemies, the boring snails and grazing fish.

The turtles are hosts of a range of specialized barnacles. In the coastal waters of Queensland the green turtles are colonized by large, thick-shelled *Chelonibia testudinaria* and *C. caretta*, which live on the scutes of their carapaces and plastrons, and by smaller *Platylepas* species which live on their scaly flippers and heads. An undescribed goose barnacle lives in the mouth and gullet of loggerhead turtles, where it is probably a commensal sharing in its

Detail of the thick-walled *Chelonibia testudinaria* which lives on green turtles, etc. (Magnetic Island)

The feather colonial hydroids provide food, a substrate for settlement, and protection for a range of symbionts such as this amphipod *Caprella*. (10m. Palm Island)

Epizoic sea anemones, half crabs, and brittle stars on a colonial hydroid. (10m. Palm Island)

This small crab perched on a sea fan sports a growth of stinging hydroids, perhaps as a protective device. Photo by Charles Arneson. (20m. Desacheo Island, Puerto Rico)

Zebra oysters *Electroma zebra* epizoic on hydroids. (10 m. Magnetic Island)

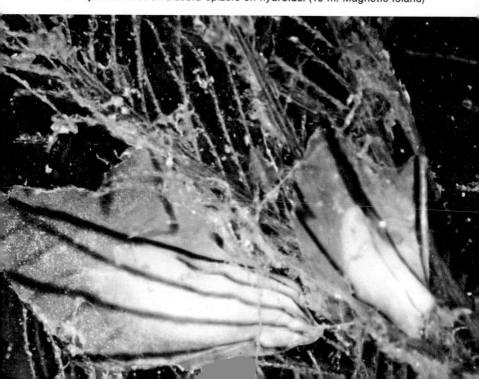

host's meal. Several other barnacles of the genera *Cylindrolepas* and *Stephanolepas* have been reported on turtles in other parts of the world.

The dugongs or sea cows of coastal northern Australia host *Platylepas hexastylos* on their backs, and the closely related *P. ophiophilus* lives on the tails of certain sea snakes. The sea snake barnacle has been reported on only rare occasions, but I have found that it occurs quite commonly on most of the dozen species of Australian sea snakes which have been examined.

Porpoises are generally free of symbiotic barnacles, but some goose barnacles of the genus *Conchoderma* have been found growing on their teeth. The larger cetaceans host many barnacles such as *Tubicinella, Coronula, Xenobalanus* (all acorn barnacles) and *Conchoderma* (a goose barnacle). *Xenobalanus* is unusual in that it has a stalk like the goose barnacles.

Conchoderma auritum is the most specialized species of its genus: as it is free from predators, it has lost its shell of protective plates. Like most barnacles it is a filter feeder, facing toward the anterior end of its whale host to take advantage of the currents. Water flows into a hood where its hairy cirri filter out plankton, and it flows out via two funnel-like tubes which resemble rabbit's ears. Other barnacles, such as the sea snake barnacle, also face anteriorly. Whale barnacle orientation has been used to trace the patterns of water flow over California gray whales.

Rocky shore barnacles rely on a strong cement, one of the most powerful known, for their adhesion. The symbiotic barnacles of mammals and reptiles rely more on mechanical methods such as ribs and roots for their attachment, probably because their hosts shed or slough their skins. The whale barnacle *Tubicinella* has a series of ridges on its shell and appears to grow by screwing down through the host's skin. *Chelonibia caretta* grows roots through the turtle's scutes into its bones. A shark barnacle (*Anelasma*) uses its roots to extract nutrients from its hosts, thus supplementing its normal filter-feeding. This trend is extended in the rhizocephalan barnacles of crabs which consist solely of these roots and reproductive organs.

One of the most perplexing aspects of these, and to some extent all other, partnerships is: how do the young manage to find their hosts in that almost infinitely vast body of water? Imagine a micro-

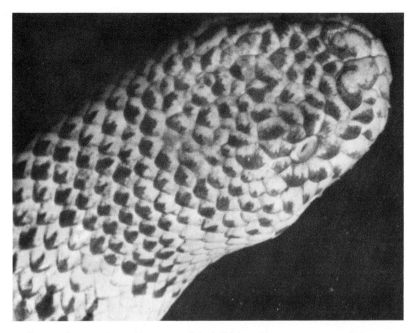

Sea snakes, such as *Aipysurus duboisii* (above) may carry a small barnacle *Platylepas ophiophilus* on their tails (below). (off Magnetic Island)

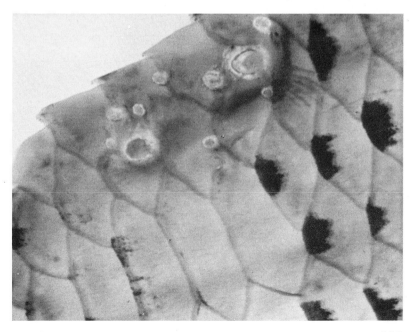

Close-up of commensal brittle stars *Ophiothela* on hydroids. (10m. Palm Island)

Mantis shrimp *Odonto-dactylus* guarding the entrance to its burrow. This burrow is also used by other animals. Photo by Allan Power.

Echiuroid worms such as this 20 cm. *Thalassema* sp. excavate burrows which may be used by other animals. Photo by Charles Arneson. (2m. Enrique Reef, Puerto Rico)

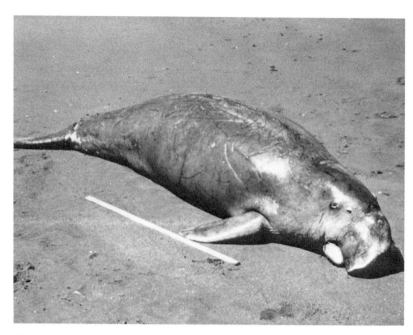

The dugongs or sea cows *Dugong dugon* host the barnacle *Platylepas hexastylos*. (Magnetic Island)

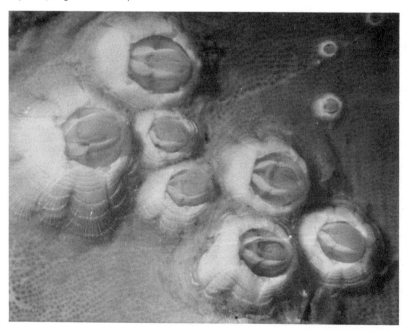

scopic barnacle larva searching for its host. Imagine one trying to attach to a whale as it surges past!

This problem seems acute in the sea snake barnacle. The adults are very small, only a few millimeters across the shell, and produce relatively few eggs for such a specialized organism. Theirs seems to be a particularly tenuous niche.

Sea snakes, even when they are common, form an insignificantly small proportion of superstrates in the open sea. The barnacle is never very common, for although I have seen over 500 individuals on a single host, only a handful are usually present on each snake. Generally few of these are sexually mature.

The odds appear to be stacked against the sea snake barnacle at each stage of its life: the planktonic larvae must avoid predation; they must locate their host (it must be far harder than 'looking for a needle in a haystack'); they must settle near other barnacles or have other barnacles settle near them (barnacles, although hermaphrodites, must cross-fertilize); they must grow to sexual maturity, cross-fertilize their eggs, brood them and liberate larvae in a limited time (sea snakes slough their skins, and most of the attached barnacles, every few weeks). How can a species persist against such odds?

Evolution of Epizoic Barnacles

Barnacles have become symbiotic for many reasons. Possibly a major one is the problem of overcrowding, for suitable substrates and superstrates are limited. On the rocks of the sea shore barnacles are often crowded shell-to-shell, sometimes several layers thick. Because suitable substrates are uncommon and competition for space is acute, the barnacles must produce many young. Many barnacles produce continuous broods of tens of thousands, resulting in millions of young per annum. Such a high fecundity also makes them well pre-adapted to symbiotic niches which are similarly difficult to locate.

Barnacles also have a high genetic plasticity, the variability demonstrated by Charles Darwin, to exploit new niches. One can envisage the evolution of a new symbiotic species. Cyprid larvae, if they cannot find a suitable surface in a certain period, will either die or settle on an unsuitable one. For a species with a high degree

The sentinal goby *Cryptocentrus* sp. stands guard as its partner the blind pistol shrimp *Synalpheus* excavates its burrow. Note the shrimp's antenna in contact with the alert fish. Photo by Roger Steene. (2m. Rabaul, New Britain)

This mutualistic association exists between different species of gobies and other alpheid shrimps. Here *Cryptocentrus koumansi* rests at the entrance to the burrow alongside the shrimp *Alpheus* sp. Photo by Dr. Gerald R. Allen. (2m. Lizard Island)

A diver examines the symbiont community living on the white gorgonian *Solenocaulon grandis* in a muddy tropical bay. (4m. near Townsville)

The scale worm *Paradyte* lives among the polyps of *Solenocaulon*.

of genetic variability some individuals may possess characteristics which enable them to settle on these 'unfavorable' surfaces and survive. This surface may be the skin of a whale or the shell of a turtle. This association would become more refined by further natural selection.

We can catch a glimpse of dynamic evolution through the sea snake, turtle and dugong barnacles (*Platylepas*). These belong to a group of barnacles once abundant on rocky shores but they later became extinct, possibly because they failed to compete successfully with another group which now dominates our shores. Thus these ancient barnacles disappeared, all except a few species which had adopted very specialized symbiotic niches. These niches are so specialized that no other species has adopted them and there is no inter-specific competition.

Unfortunately their future is no longer secure. This time their competitor is man, who has over-hunted turtles, dugongs and manatees. These hosts and those of the whale barnacles are now endangered. But even if these reptiles and mammals are saved from extinction it may be too late for the barnacles, which are finely tuned to their hosts' biologies. As the hosts become less common the minute larvae have less chance of finding them and fewer adults reach maturity. Thus in turn fewer eggs are produced. Perhaps during our lifetime many of these symbiotic barnacles will slide, unnoticed and unmourned, into extinction.

PART III. HOSTS FOR SYMBIOSIS
Chapter 7. Molluscs and Ascidians

In the previous chapters certain organisms appear repeatedly as symbionts, while others appear equally often as hosts. The organisms which are particularly prone to exploitation by symbionts are often sessile, complex in body form and plankton feeders. Some are colonial and capable of considerable regeneration of lost parts, while others have certain defenses which may be used by refugees. Such hosts include the sponges, the coelenterates (especially the corals, jellyfishes and anemones), the echinoderms, the molluscs (mainly the bivalves) and the ascidians.

The symbionts of corals, jellyfishes, hydroids and sponges have been discussed earlier. This chapter and the next outline the symbionts of the other common hosts.

SYMBIONTS OF MOLLUSCS

The phylum Mollusca is the second largest in the animal kingdom, with about one hundred thousand species. (The largest phylum, boosted by the insects, is the Arthropoda). The molluscs, primarily marine organisms, are made up of six classes: Monoplacophora (a small deep sea group); Polyplacophora (the chitons of rocky shores); Gastropoda (snails); Scaphopoda (tusk shells); Pelecypoda (bivalves); and Cephalopoda (octopuses and squids). The gastropods and bivalves are the most common, com-

Even when the white gorgonian dies it is still important as a substrate for organisms seeking a surface for settlement. Sea anemones, sponges, and hydroids encrust the horny remains of a *Solenocaulon*.

The allied cowry *Primovula rutherfordiana* and two small crustaceans on the tip of *Solenocaulon*. Note the small white processes on the snail's mantle which resemble the host's polyps.

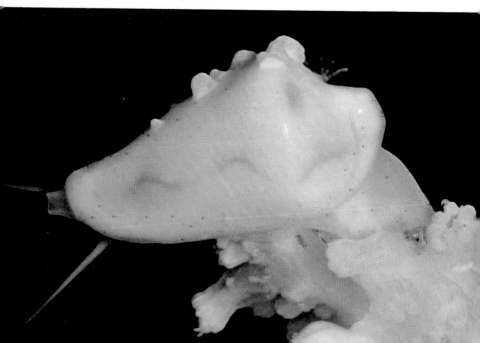

The brittlestar *Ophiopsammium* cf *semperi* lives wrapped around the branches of *Solenocaulon* (below). It is shown above removed from its host.

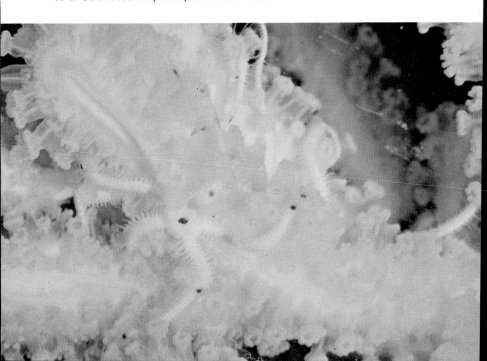

prising over 90% of all molluscs. Both, but particularly the bivalves, are common hosts for symbiosis.

The snails are mainly marine, but some are freshwater and terrestrial. They range in size from millimeter-long interstitial species to the large predatory tritons over one half meter in length. Most gastropods have a single shell used to protect their visceral masses, but this is reduced or absent in certain groups. Likewise most have a relatively large locomotory organ called the foot.

The marine snails occupy a large range of niches in most marine ecosystems. They are especially common in intertidal areas and on hard and soft bottoms. Snails may be herbivores, rasping algae and diatoms from rocks, or carnivores, with specialized feeding habits. A few feed on plankton.

Snails are hosts for a variety of refugees because of their large semi-enclosed mantle cavities which are usually protected by solid shells. They have a soft body covered with mucus which may be stolen by commensals, and they have a few defenses to repel invaders.

The bivalves are filter-feeders with, as their name implies, two shells or valves to protect them. Like the gastropods they are predominantly marine but include some freshwater members. Bivalves are even more heavily exploited than the gastropods due no doubt to the readily available food in their large protective mantle cavities. These molluscs set up strong feeding and respiratory currents, pumping plankton-containing water through sieve-like gills where edible material is trapped in mucus and eaten. Many commensals pirate this food before it reaches their hosts' mouths.

Microorganisms

Very little is known of the parasitic microorganisms which undoubtedly infest the invertebrates. It is known that the commercially cultured bivalves are prone to disastrous epidemics caused by pathogenic bacteria, fungi and protozoans. Bacteria may produce tissue necrosis and abscesses; fungi may cause diseases of the basement membranes, gonads and shells; and protozoans, mainly sporozoans such as plasmodia, infect gills, connective tissue, the gut and gonads.

Algae

A few of the gastropods culture symbiotic dinoflagellates known as zooxanthellae. A nudibranch, *Aeolidiella glauca*, acquires zooxanthellae from its sea anemone prey and cultures them in its own tissues, but the role of the algae in the new host is not known. Some of the sacoglossan molluscs take chloroplasts from the cells of the algae they consume and incorporate them into their own cells. There the chloroplasts continue to photosynthesize.

The bivalves, notably the giant clams, more often culture algae in their tissues. In *Tridacna* zooxanthellae are cultured in special organs scattered over the surfaces of their expanded and highly colorful mantles. Small lenses in these organs concentrate sunlight, thus increasing the photosynthesis of the zooxanthellae cultured in them. The algae receive nutrients and protection from their hosts while the clams benefit by having their wastes removed and supplement their diets by eating zooxanthellae. Specialized blood cells harvest the cultured algae by engulfing them and take them to the clams' digestive glands. Thus symbiosis has possibly enabled the *Tridacna* to reach their great size—up to 1.5 m. in length and 200 kg. in weight.

Sponges

The sponge *Cliona* bores galleries in mollusc shells as well as in living and dead corals. Often the molluscs repair these holes by depositing conchiolin, the organic material of molluscan shells, as sponge tissue which penetrates the shell may damage the soft parts. Epithelial and connective tissue may be dissolved, sometimes resulting in the death of the host. The method of excavation has been discussed earlier.

Coelenterates

Certain of the hydractinid hydrozoans live as epizoites on certain of the mud or basket snails (Nassariidae). In the temperate waters of the Indo-Pacific the hydroid *Hydractinia* lives on the shell of the small sand snail *Parcanassa*. In the tropics of the Pacific this symbiosis is represented by the hydroid *Stylactella*, which lives on a similar sand snail, *Niotha*.

Fishes hover around drifting objects as a source of shelter and/or food. These young *Seriola zonata* temporarily shelter beneath this object until something better comes along. Photo by R. Robert Abrams.

Shelter in open water is limited and many occupants have entered into associations with large or dangerous species. Note the reef fish, fusiliers *(Caesio),* eating a planktonic ctenophore. (Lodestone Reef)

Portuguese Man-O'-War fish *Nomeus* shelters with immunity among the deadly tentacles of the siphonophore *Physalia*. Other small fishes coming in contact with the nematocysts are stung and eaten. Photo by Charles Arneson.

A cleaned shell of the dogwhelk *Niotha* alongside a shell covered by hydroids. (Intertidal, Magnetic Island)

The hydroid is able to exploit the soft sand and mudflats because the snail prevents it from being buried by sediment. It may also partake in its hosts' meals, for the basket snails are scavengers; when they rasp their food, particles of it cloud the water. The hydroids have been observed taking fragments of food. The association is probably obligatory as these snails are rarely seen without their partners.

Hydroids may similarly form associations with the bivalves which live on soft bottoms. In Scotland the hydroid or sessile stage of the jellyfish *Neoturris* lives on the edges of a nut-shell (*Nucula*) which burrows into fine mud. These hydroids are well able to live beneath the mud, but a related hydroid (*Leuckartiara*) which lives on another nut-shell (*Nuculana*) is less tolerant to mud and grows to the surface on long stems.

Worms

The parasitic trematodes and cestodes and the role of snails as intermediate hosts have been discussed in an earlier chapter in some detail. It has been said that every species of marine snail serves as host for some species of trematode and many snail species host more than one species of parasite. The life cycles of these parasites are very complex and the intermediate hosts are vital for increasing their reproductive capacities by a process of larval multiplication.

Hosts of these parasites include many of the fishes, the reptiles and sea birds. Man may occasionally be a host of marine blood flukes. These may burrow into the bodies of bathers and fishermen, but often the victim's dermal tissue reacts to encyst the foreign protein to produce the irritating lesions of cercarial dermatitis or 'clam-diggers itch'. The freshwater flukes are among man's greatest enemies, for diseases such as bilharzia continue unchecked in the Third World. Some of the bivalves also act as intermediate hosts for trematodes and cestodes. Larvae of the cestode *Tylocephalum* are thought to induce pearl formation in the pearl oysters.

The keyhole limpet *Diodora* of North America often hosts the polynoid worm *Arctonoe* in its mantle cavity. The worm, sometimes as long as the host itself, may defend its host from its enemy, the predatory starfish *Pisaster*. If a starfish approaches the keyhole limpet the worm will slide out from under the mantle and bite one of the tubefeet of the attacking starfish. This generally drives it away.

Crustaceans

Many copepods parasitize the gastropods and bivalves. Some live on their soft body surfaces and eat mucus and epithelial tissue but others live in the protected mantle cavities, on the gills, in the intestines and in the oviducts.

The pinnotherid or pea crabs are much better known. As their name implies these have a pea-shaped body and weak, ineffectual legs. The cuticles of the females are reduced in thickness for they are effectively protected by their hosts. They have also lost some of their sensory organs. The pea crabs were well known to the ancient Greeks and were possibly the first marine symbionts ever described.

Eyebrowfish *Psenes pellucidus* sheltering among the tentacles of the jelly-fish *Dactylometra pacificus*. Photo by Y. Takemura and K. Suzuki from Seashore Life in Japan.

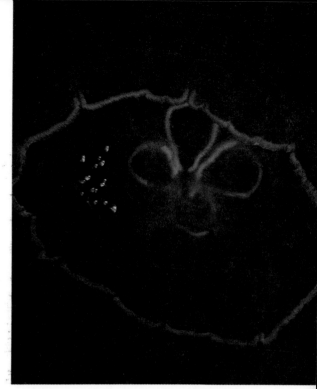

The small silvery objects within the bell of this jellyfish *Aurelia aurita* are young carangids *Chloroscombrus chrysurus.* Photo by Charles Arneson. (Puerto Rico)

A close-up of the small school of *Chloroscombrus* hiding in the bell cavity. Photo by Charles Arneson. (Puerto Rico)

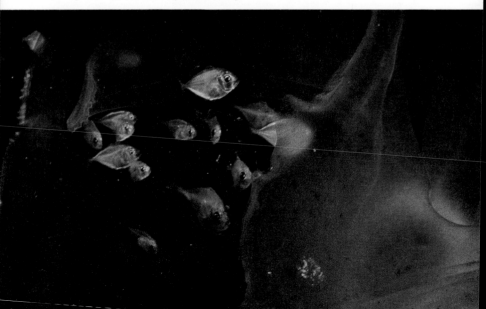

Many species of pea crabs live in gastropods and bivalves as well as in worms, sea cucumbers and sea squirts. Their mode of feeding varies: some are filter feeders straining plankton with setose appendages; others, after an initial phase of filter-feeding, steal food from their host's gills.

The feeding behavior of the European *Pinnotheres pisum* has been studied by cutting holes in the shells of mussels and inserting transparent windows. This crab was seen to pick up food strings from the edges of the mussel's gills using specialized nippers or chelipeds.

Pea crabs undergo a planktonic phase made up of a varying number of zoea and megalopa stages. The first crab stage invades the particular host and the males and females mate shortly after. The female then molts and assumes her familiar soft-shelled appearance while the tiny male leaves her in search of another female. However, if the females of the oyster crab (*P. ostreum*) are starved, they then change into males and become free-living again. This is an intriguing way of maintaining the reproductive capacity of the individual under adverse conditions. Adult females expend far more energy in egg production than males use in sperm production, so if the resources are not available for the female it takes the alternate and less energetic role of the male.

P. pisum has a more complex life cycle than other pea crabs. After a number of planktonic stages the small crabs invade the bivalve *Spisula*, but after growing to a larger size they leave to seek out a mussel in which to complete their development.

Most of the pea crabs are regarded as commensals, but some cause their hosts to lose condition, thus affecting egg production. These species are generally termed parasites. Mostly the damage to the host is slight, but sometimes a crab's dactyls (claws) may induce the formation of a tumor-like swelling in the host's mantle.

Only single females of most pinnotherids live in each host, indicating that the crabs rigidly control their numbers by excluding others of their kind. This prevents the over-exploitation of the host.

The large bivalves of the Indo-Pacific often host a pair of small shrimps belonging to the subfamily Pontoniinae. Like the pea crabs these shrimps live in their hosts' mantle cavities, probably as commensals. The pen shell shrimp, *Anchistus custos*, for example,

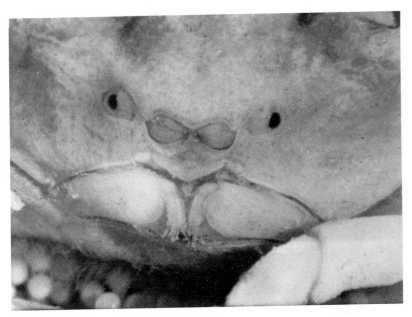

Detail of the clam crab *Xanthasia.* Note its reduced eyes.

has a modified pair of walking appendages equipped with fine fringing bristles or setae. The shrimp uses these appendages to brush mucus and trapped plankton from its host's gills. This material is then removed by wiping the brushes through another pair of appendages equipped with a special groove and is eaten.

The bivalve shrimps are very specific in their host preferences. In the Indo-Pacific the pearl oyster shrimp, *Conchodytes meleagrinae,* lives in the mantle cavity of the black-lip pearl oyster, *Pinctada margaritifera,* or occasionally in the cloacas of some sea cucumbers. The latter may be alternative hosts if the shrimps cannot find an unoccupied *Pinctada.* A closely related shrimp, *C. tridacnae,* lives in the *Tridacna* clams and a third, *C. biunguiculatus,* lives in pen shells (*Pinna*). These shrimps are thin-shelled with swollen bodies and weak walking legs. The females are much larger than the males, which have distinctively enlarged nippers. The function of the nippers in such a protected habitat is unknown, but it has been suggested that they are used to drive away or dispatch newcomers as only one pair of shrimps occupies each host.

Above and below. Juvenile herring trevally *Caranx kalla* sheltering near the jellyfish *Cyanea capillata*. When alarmed the fish retreat into the tentacles. (Magnetic Island)

The pelagic nudibranch *Glaucus atlanticus* eats siphonophores such as *Physalia, Porpita* (above), and *Velella* (below), and uses the stinging nematoblasts for its own protection. Photo above by Charles Arneson (La Parguera, Puerto Rico); photo below by the author (Evans Head, New South Wales)

Infestation rates may be very high. In certain rich beds of coral reef pearl oysters I have found that as many as 90% of the mature oysters contain a pair of shrimps. Immature oysters may only contain one shrimp or sometimes none. Of mature oysters in inshore areas only 20-30% contain shrimps, indicating that the symbiont is less tolerant of salinity, temperature and/or turbidity changes than the host. The high incidence of infestation of coral reef oysters indicates that the shrimps are very successful in locating their hosts, that the established pair deters any newcomers, and that the success of the oyster is little affected by its inseparable partner.

Occasionally these shrimps share their hosts with pairs of *Anchistus* and *Paranchistus*, less modified shrimps also showing a high degree of host specificity. *A. custos* lives in *Pinna*; *A. demani* lives in *Tridacna maxima*; *A. miersi* lives in *T. squamosa*; and *A. pectinus* lives in scallops. The related *P. ornatus* lives in the pen shell *Atrina*, and *P. biunguiculatus*, the largest of all the commensal shrimps, appropriately lives in the giant clam *T. gigas*.

In the Mediterranean and tropical Atlantic the genus *Pontonia* has filled the niche of bivalve commensal. *P. margarita* lives in pearl oysters and *P. pinnae*, *P. domestica* and *P. mexicana* live in various pen shells, *Pinna*.

Only one shrimp, the superbly colored emperor shrimp, *Periclimenes imperator*, lives with the gastropods. Its host, appropriately enough, is one of the most beautiful creatures in the sea, the fabulous Spanish dancer nudibranch (*Hexabranchus*). When *Hexabranchus* swims it swirls and pirouettes, flashing blood red and purple folds of mantle as a flamenco dancer swirls her skirts.

The emperor shrimp feeds on it s host's mucus, on detritus which adheres to this mucus and on materials it picks up from the surrounding substrate as the host crawls along. The Spanish dancer is uncommon and the young shrimps must experience difficulties in locating it. If they fail to find an unoccupied host they may pick another species of nudibranch or even the sea cucumber *Stichopus*. However, they cannot grow to maturity on the latter—perhaps they use it as an interim host until something more suitable comes along.

On the Great Barrier Reef an alternative host to the Spanish dancer is the red nudibranch *Ceratosoma*. I once kept this nudibranch and a pair of shrimps in an aquarium for some months but the shrimps ultimately ate their host, probably because they required food not present in the aquarium and were forced to rely too much on the host for food.

Molluscs

A number of molluscs live as symbionts on their fellow molluscs. These include parasites such as the pyramid shells which live on large bivalves and suck their body fluids through long proboscises. *Odostomia* for example, although free-living, spends much of its time on its host (*Pinctada*) sucking blood from the oyster's mantle.

Several small limpet-like snails (*Crepidula, Saptadonta, Sabia* and *Hipponix*) live permanently attached to large coral reef snails such as *Turbo, Trochus* and *Lambis. Hipponix* excavates a pit in the host's shell near its exhalent siphon and secretes a limy base onto which its shell snugly fits. It feeds on its host's feces and on detritus which settles near it.

An epizoic hipponicid snail attached to the spider stromb *Lambis truncata*. The snail eats its host's feces and other materials which settle near by. (10m. Grubb Reef)

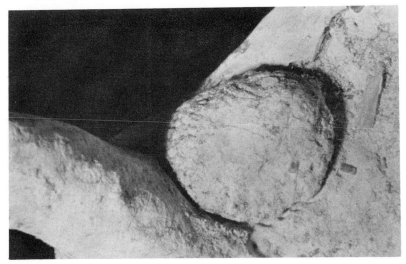

Glaucilla marginata, one of the pelagic nudibranchs. Photo by T. E. Thompson.

This juvenile pomfret *Parastromateus niger* was forced onto the tentacles of *Cyanea* while being captured. Note the blue sting marks. This fish later died of its injuries while the trevallys *Alepes* in the background remained unaffected by contact with the tentacles. (Magnetic Island)

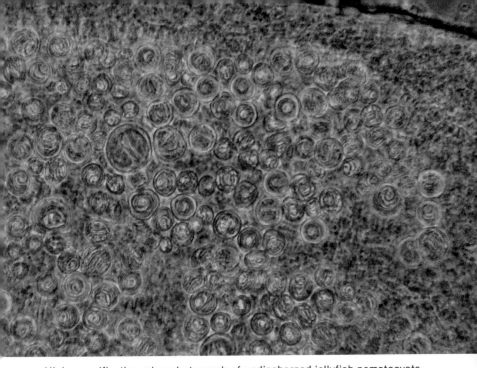

High magnification microphotograph of undischarged jellyfish nematocysts in the tip of a ceras from the pelagic nudibranch *Glaucilla.* Photo by T.E. Thompson.

Discharged nematocysts from the cerata of the nudibranch *Aeolidia papillosa.* Photo by T.E. Thompson.

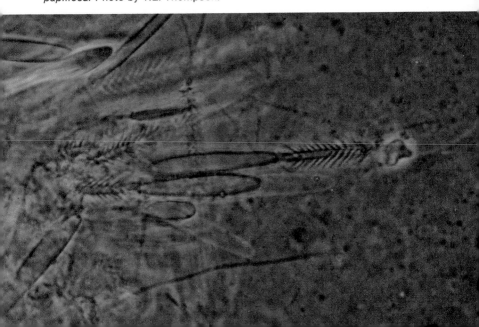

Cluster of epizoic cap shells *Capulus* on a snail shell. (10m. Wheeler Reef)

Although many mussels, oysters and other attached bivalves often foul the shells of other molluscs, few have a regular association with a particular host. An exception is the epizoic moon mussel *Ciboticola* which lives on the exterior of certain *Arca* shells not much larger than itself. The host burrows into the soft sediment of tropical mud flats but protrudes its siphons into the water above. The mussel, although it has limited powers of locomotion, needs a solid surface on which to attach to avoid being smothered, so it lives on the arc shells which are better adapted to life on soft bottoms. Its shells are adapted to life on its partner: one surface of each valve is concave so it snugly fits over the rounded shell of the host.

The boring mussel *Lithophaga* often bores cavities in the larger molluscs but the associations are not obligatory. Pearl oysters may produce blister pearls as they repair the holes made by these borers. If silt and sand leak into the oysters they may secrete a layer of pearly nacre over the sediment, thus creating a hollow blister pearl.

Fishes

The conchfishes *Astrapogon stellatus* and *A. puncticulatus* shelter in the large mantle cavities of the giant conchs (*Strombus*) of the tropical Atlantic. Like other cardinalfishes the conchfishes are nocturnal and at night leave their hosts to forage.

Pearlfishes or messmates of the family Carapidae similarly shelter in the mantle cavities of the large bivalves and in the gut and respiratory cavities of sea cucumbers. *Carapus homei* primarily lives in sea cucumbers but may shelter in giant clams and sea squirts. *Onuxodon margaritiferae*, a large-fanged pearlfish, shelters only in pearl oysters.

The pearlfishes are quite bizarre in appearance. They are of finger-length, long and slender with a filamentous tail. They lack pigment and are remarkably transparent. It is possible to see the internal organs such as the brains, larger nerves, heart, blood vessels with streaming blood, the entire alimentary canal, and the vertebrae and ribs through the body.

The pearlfish *Onuxodon margaritiferae* is almost completely transparent. It hides in its host during the day and feeds outside at night.

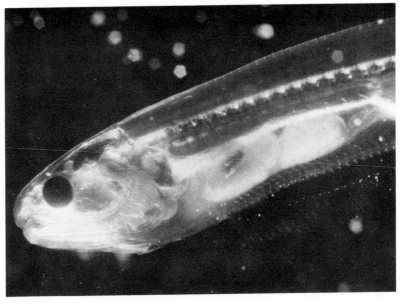

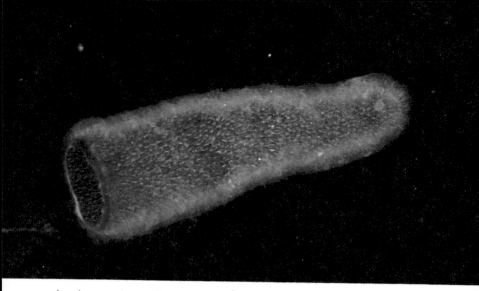

Luminous salps such as this *Pyrosoma* may become parasitized by a small amphipod *Phronima*. The amphipod gains food, protection, and a means of dispersal. Photo by Charles Arneson. (12m. Desecheo Island, Puerto Rico)

The purple bubble snail *Janthina* provides a substrate for some barnacles of the genus *Lepas*. Photo by Charles Arneson. (La Parguera, Puerto Rico)

Shell-less stalked goose barnacle *Alepas* cf *parasita* removed from its host jellyfish *Cyanea*. (Magnetic Island)

Lepas anserifera feeding as they drift along attached to a piece of old wood. Photo by Aaron Norman.

Pearlfishes feed at night in the outside waters and shelter during the day in their hosts. An interesting adaptation to their way of life is the anterior migration of the anus. They do not defecate in their hosts; by having the anus near their head they need not protrude much of their body from shelter when defecating. Evolution for hygienic reasons!

Pearlfishes have been found entombed in the pearly nacre of their pearl oyster hosts. It is possible that these fishes burrowed between their host's mantle and shell. Having become trapped, their dead bodies might then have been isolated by the deposition of nacre.

SYMBIONTS OF ASCIDIANS

The colonial ascidians and solitary ascidians, the sea squirts, are a minor group of soft-bodied marine animals closely related to the vertebrates. They are here included with the molluscs because the sea squirts occupy a niche similar to that of the bivalves and offer similar benefits to symbionts.

Like the bivalves, the sea squirts are filter feeders. They pump sea water in via an inhalent siphon and filter it through a net known as the branchial basket which lies in a cavity, the atrial cavity. The food trapped on the branchial basket is collected by ciliary action and taken to the sea squirt's esophagus attached to a continuous cord of mucus. A tough case or test encloses the sea squirt's soft parts.

Sea squirts host many commensals, no doubt due to their large enclosed branchial and atrial cavities, the filter-feeding habit which brings in continuous supplies of food and water currents, the ease of entry and departure via the inhalent and exhalent siphons and the lack of defense mechanisms to repel invaders.

Algae

A number of tropical didemnid ascidians host algae in their cloacal cavities and tests. *Trididemnum* has closely packed green algae (zoochlorellae) in lacunae in the test which surrounds the colony but it is not known what benefits, if any, the host receives. However, a number of ascidians culture algae in special organs

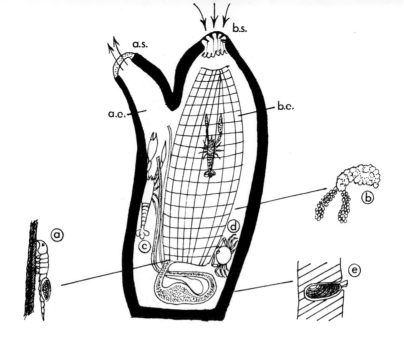

Body plan of an ascidian sea squirt, showing branchial cavity (b.c.), atrial cavity (a.c.), buccal siphon (b.s.), and atrial siphon (a.s.). The positions of: (a) an esophageal copepod *Ascidicola rosea;* (b) a body wall copepod *Gonophysema;* (c) the shrimp *Pontonia;* (d) a pea crab *Pinnotheres;* and (e) a bivalve *Musculus,* are shown. (From various sources)

which have no other function, indicating that the symbiosis gives the host some selective advantages. The larvae of *Diplosoma virens* even have pouches which are used to transport their symbiotic algae. The algae are green (zoochlorellae), blue-green or, as recent research indicates, intermediates between the greens and blue-greens.

Crustaceans

The best known commensals of the ascidians are the copepods. Species of one family, the Notodelphyidae, are almost exclusively associated with ascidians. To date well over 100 species associated with ascidians have been described and undoubtedly many more remain to be discovered. As well as these, many species of the Lichomolgidae, Archinotodelphyidae, Ascidicolidae and Enterocolidae also live in ascidians.

Generations of sea snake barnacles last only as long as their host's skin—only about one month. *Platylepas* is shown here adhering to the shed skin of the snake *Lapemis hardwickii.* (off West Irian)

The green turtles *Chelonia mydas* host several barnacles, for example the large *Chelonibia testudinaria.* The lumps on the carapace are made by a boring barnacle. (Raine Island)

Detail of the mantle of *Tridacna maxima*. The color is due to the symbiotic algae. (3m. Lodestone Reef)

The giant clams *Tridacna* farm symbiotic algae in their colorful mantles and shelter a number of crabs, shrimps, and other symbionts within. (Intertidal, Percy Island)

Little is known of the effect of these symbionts, but as infestations are often particularly high they must have some influence. Single sea squirts have been found to contain dozens of copepods of several different species as well as numbers of amphipods, polychaetes and crabs.

Many of the copepods live as commensals. *Ascidicola* lives inside the transparent sea squirt *Corella* and may be seen clinging to the mucus cord which brings the planktonic food to the esophagus. The cord is constantly moving like an endless conveyor belt so the copepod must periodically readjust its position by crawling up a few millimeters. Its eggs are deposited in the host's gut and are expelled with the feces via the exhalent current. The life cycle of *Ascidicola* is closely linked to that of the host for it produces its larvae when a newly settled generation of its host is established.

Certain of the copepods have adopted a parasitic existence in cysts within tissues, blood vessels or the body wall. For example *Gonophysema*, a copepod which lacks a mouth and gut, lives bathed in blood in its host's body wall. Its two egg sacs protrude into the ascidian's body cavity.

Close-up of the ascidian shrimp *Pontonia katoi*. Note its small rostrum; most symbiotic shrimps have reduced spines, an adaptation to life in cramped quarters.

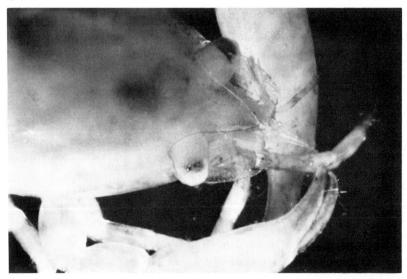

A number of amphipods also live in the branchial sacs of ascidians. Some of these, for example the blood-red *Paraleucothoe*, almost fill their hosts' branchial basket.

Shrimps of the genus *Pontonia* live inside some of the sea squirts. Like the amphipods they may almost entirely fill their host, and it is generally accepted that they enter as larvae and as they grow become trapped within. However, at least some of these shrimps may be able to leave their prisons at will for I have found that *P. katoi* and *P. sibogae* are expert swimmers and crawlers when they are removed from their hosts and have well developed chromatophores for camouflage. One might expect a life-long prisoner to have weak legs and uropods and reduced chromatophores.

The sea squirts also host a few of the pea crabs. *Pinnotheres pinnotheres* lives in the atrial cavity of European ascidians and filters food from the water current produced by its host.

As one might expect, the interiors of the sea squirts can get very crowded and there appears to be inter- and intra-specific competition among symbionts to reduce numbers. Many of the larger species restrict their numbers to one or two individuals, probably through territoriality. There is evidence that the pea crab *Pinnotheres* will tolerate the amphipod *Leucothoe*, but neither will allow the shrimp *Pontonia flavomaculata* to become established.

Ascidians may be used for camouflage by certain crabs. The sponge or dromid crabs may carve a cap of ascidians to wear, and the majid or decorator crabs may plant pieces of colonial ascidians on their carapaces and legs.

Worms, Mulloscs

A number of small worms live in the ascidians. Polychaetes have been found in the cloacal cavities and a nemertine worm may live in the branchial or atrial cavities.

Several small snails and bivalves may embed themselves in ascidian tests. The bivalve *Musculus* burrows into the tests of *Ascidia* to gain shelter. Similarly, a mussel may attach to the tests with its thread-like byssus, ultimately embedding itself. Over 40 individual bivalves have been found in the test of a single ascidian. The mussels apparently recognize the host by some chemical fac-

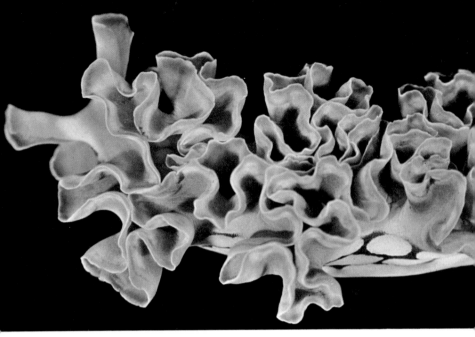

The nudibranch *Tridachia crispata* extracts chloroplasts from the cells of the algae on which it feeds and incorporates them in its own body wall. Photo by T.E. Thompson. (Caribbean)

Epizoic hydroids *Stylactella niotha* on the dogwhelk *Niotha albescens*, seen here feeding on a dead oyster. The hydroids camouflage the snail and may share in some of its food. (Intertidal, Lizard Island)

The intertidal snails such as these *Neritina* and *Cerithium* are the intermediate hosts for many bird and fish flukes. (Intertidal, Magnetic Island)

The keyhole limpet (*Diodora*) of North America often hosts the polynoid worm *Arctonoe*. This worm lives in the mantle cavity and may defend its host from predatory starfish. Photo of *Diodora inaequalis* by Alex Kerstitch. (Isla San Nicholas, Mexico)

tor in the test for they respond to pieces of dead test as readily as to those of living animals.

SYMBIOLOGY OF MOLLUSCS AND ASCIDIANS: CONCLUSION

Molluscs and ascidians, although unrelated, have many similarities. Both are soft-bodied but are protected by calcareous shells or strong tests. Both have semi-enclosed cavities protected by shells or tests, and both pump water into these cavities. The bivalves and ascidians occupy similar niches. Both are filter-feeders pumping water through a mucus-covered net which strains out planktonic food. This food is then transported to their mouths stuck to mucus.

As might be expected, the well protected cavities offer protection to organisms seeking shelter; water currents are utilized for respiration or for filter-feeding; and mucus and trapped food are stolen by commensals.

A number of symbionts, for example the pea crabs and the shrimps, are common to each group. Probably these evolved firstly in the molluscs and exploited the ascidians secondarily. Like many of the other symbionts described in this book, a number of mollusc and ascidian symbionts show similar specializations to their ways of life and many avoid overtaxing their host's resources by restricting their own numbers.

Chapter 8 Echinoderms

Like the coelenterates, sponges, molluscs and ascidians, the echinoderms are exploited by a range of organisms seeking food and protection.

The phylum Echinodermata is exclusively marine and benthic and is represented in habitats from the lower intertidal to the ocean deeps. It is composed of five classes: Asteroidea (starfishes); Echinoidea (urchins); Crinoidea (featherstars); Ophiuroidea (brittlestars); and Holothuroidea (sea cucumbers). These are all characterized by their radial symmetry of five divisions around a central disc, their calcareous plates and their water vascular system (a kind of hydraulic system which provides turgidity to their many tube feet, appendages used in locomotion).

Echinodermata means 'spiny skinned' and refers to the calcareous plates which form an exoskeleton in all classes other than the sea cucumbers. The name also gives a clue as to why they are popular refuges to many organisms seeking shelter: their complex body shapes and defensive spines provide many protected habitats.

Although the echinoderms attract many symbionts, the phylum is unique among the major phyla in that no members have taken the step into parasitism. Whey they have failed to adopt parasitism is uncertain but, as they are among the most complex of the invertebrates, their body plan may be too restrictive to undergo the modifications necessitated by that way of life.

SYMBIONTS OF STARFISHES

The Asteroidea or starfishes are the best known of all the echinoderms. They have a distinctive star-shaped appearance,

Commensal *Anchistus custos* on the shell of their host, the bivalve *Atrina*. (3m. Orpheus Island)

Barnacles and other animals attached to the base of the siphons of the clam *Entodesmia saxicola*. Photo by T.E. Thompson. (NE Pacific)

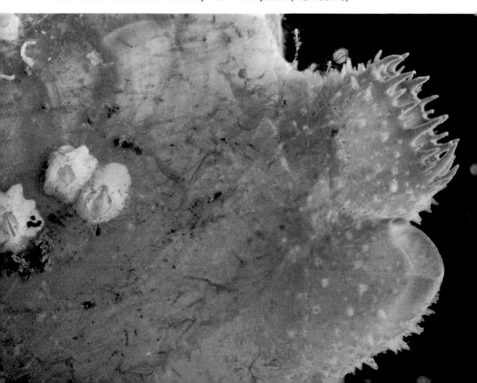

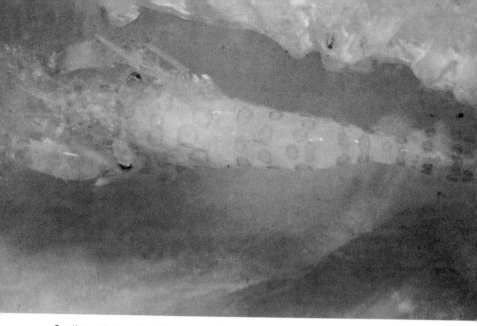

Scallop shrimp *Anchistus pectinus* on the gills of the saucer scallop *Amusium balloti*. (30m., off Wheeler Reef)

Close-up of the chromatophores of the scallop shrimp.

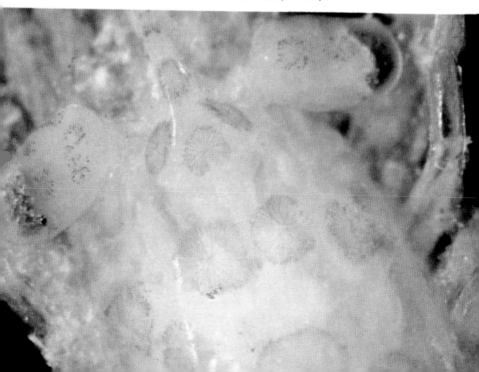

generally with five points or arms, and their bodies are protected by a rigid exoskeleton of articulating plates. At the center of the body, on the lower or oral surface, lies a mouth, radiating from which are five slits called ambulacral grooves. A multitude of minute tube feet, usually grouped in pairs and often capped with suckers, protrude from these slits. In locomotion the tube feet grip the substrate and push the body forward then release the grip of the sucker and take another step forward, rather like columns of soldiers marching in step.

Many starfishes feed on bivalves, using their robot-like arms and hydraulic system to tear the shells apart with a slow but incredible force. They then extrude their pouchy stomachs from their mouths and digest their prey's soft parts externally.

Crown-of-Thorns Starfish

The most unusual starfish of all feeds on stony coral. The crown-of-thorns starfish, *Acanthaster planci*, envelops the coral with its sixteen or so arms and extrudes its large stomach to cover an area of polyps. Enzymes are released and the coral's soft parts are liquefied and absorbed, leaving only the white calcareous skeleton.

Stories of hordes of large and spiny coral-eating starfish which laid waste to bountiful coral reefs were initially treated with disbelief by the public and biologists alike. Until recently it was thought that starfishes ate only molluscs and that the corals' nematoblasts effectively repelled most would-be predators. The culprit, the crown-of-thorns starfish, was also once considered to be quite a rarity.

Like locusts, plagues of these starfish appeared almost overnight in widely separated areas of the Indo-Pacific. Many reefs were devastated and certain scientists held fears for the entire Great Barrier Reef and the coral reefs and atolls of Oceania.

A number of governments and research institutions of the Indo-Pacific instigated research projects on the starfish and its control. Interest was sparked in the symbiology of the starfishes and the echinoderms in general, for it was hoped that a biological control might be found to halt the depredation. These studies doubled the known symbionts of starfish.

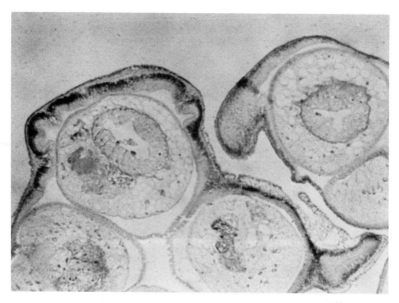

Cross-section through a lobe of the hepatic caecum of the crown-of-thorns starfish packed with the parasitic turbellarian *Pterastericola vivipara*. Photo by L. Cannon.

Parasitic Helminths

The crown-of-thorns and certain other starfishes are parasitized by turbellarians (*Pterastericola*). These worms feed on epithelial tissue of the hepatic caecae of the gut. Their biology was investigated in the hope that they might be used as a control for the crown-of-thorns, but they were found to be 'good' parasites, causing little apparent damage to their host.

Polychaete Worms

A number of polynoid scale worms live on the body surface of the starfish. These are one to two centimeters in length, dorsoventrally flattened for better adhesion and covered with two rows of scales or elytra. They are invariably well camouflaged to accurately match their host's pigmentation, and their paddles or parapodia may resemble their host's tube feet or spines. *Hololepidella* is a well camouflaged symbiont of the crown-of-thorns. It too does little damage to its host as it probably feeds mainly on mucus.

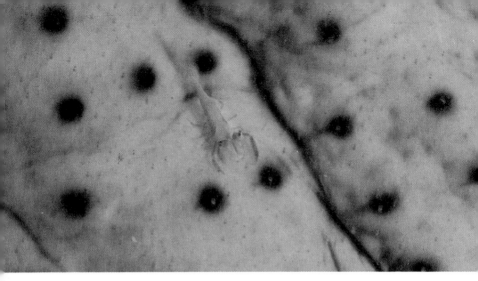

If a young emperor shrimp cannot find a suitable host it may settle on one of the common sea cucumbers of the genus *Stichopus*. In later life it may transfer to a nudibranch if one comes by. (3m., Palm Island)

Copepods under the gills of the nudibranch *Casella atromarginata*. (3m., Palm Island)

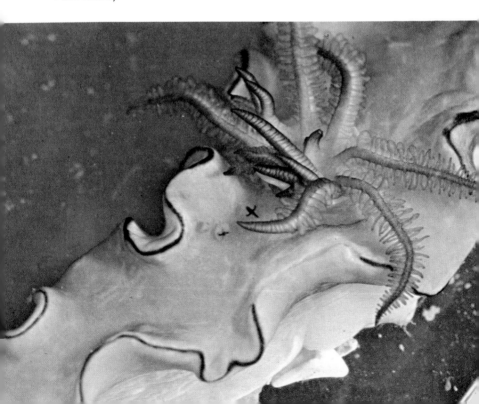

Periclimenes imperator feeding on material on a coral. This Spanish dancer is the shrimp's usual host. Photo by Aaron Norman.

Close-up of the above association. *Periclimenes imperator* is perfectly camouflaged on the Spanish dancer. Photo by Aaron Norman.

A long-bodied nereid worm may live in the ambulacral grooves of certain starfishes. These receive shelter but may travel along the groove to the host's mouth and steal a meal.

Crustaceans

Copepods often live on the surfaces of starfish. The beautiful cobalt blue starfish of the tropical Pacific, *Linckia laevigata*, hosts many minute copepods, *Linchiomolgus*, which forage on its body surface. These copepods exactly match their host's distinctive color and would be completely invisible but for the unpigmented white egg sacs which trail from the tails of the ovigerous females.

Another *Linckia*, this one from the Red Sea, is variably colored and its copepods are similarly variable. Some individual hosts are multicolored and are populated by copepods of matching colors. A copepod of one color cannot stray onto those parts of its hosts which may be a different color or it will immediately become apparent.

Two species of shrimps, *Periclimenes soror* and the larger *P.*

The starfish copepod *Linchiomolgus coeruleus* is the identical deep blue of its host *Linckia laevigata*. Only its unpigmented egg sacs spoil the camouflage. (2m., Lodestone Reef)

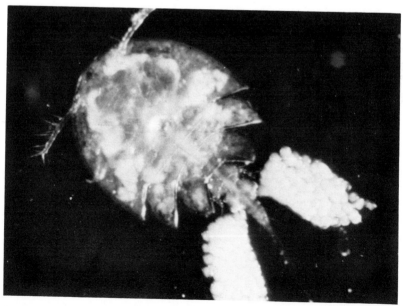

Ectoparasitic snail *Thyca crystallinus* on the starfish *Linckia laevigata*. (3m., Lizard Island)

noverca, live on starfishes of the Indo-Pacific. The former often live among the spines of the crown-of-thorns and may bear a spine-like stripe along their dorsal surfaces to perfect their camouflage. On their other hosts these shrimps generally hide on the lower (oral) surfaces, but those inhabiting the crown-of-thorns wander all over it. *P. noverca* is territorial with only one or two shrimps on each host, but I have seen 34 *P. soror* living on a single giant pin-cushion star (*Culcita*). The feeding habits of these shrimps are not known, but it is assumed that they feed on mucus and perhaps epithelial tissues. *P. soror* living on the cobalt-blue *Linckia* may in-corporate their host's distinctive pigment in their own bodies.

The only other crustaceans to parasitize the starfishes are the dendrogastrid barnacles. These highly degenerate ascothoracid barnacles, which consist mainly of an ovary, live bathed in their hosts' body fluids. Crown-of-thorns may be parasitized but the parasite does little damage.

Gastropods

Considering the large number of gastropod species, very few are parasitic. Most of those which are—four families—are parasites of the echinoderms. These range from the relatively unspecialized eulimids, which crawl about from host to host to feed on their body fluids, to the highly modified and degenerate endoparasites.

Large bivalves with enclosed mantle cavities accessible via the inhalent and exhalent siphons attract many refugees. Trapped plankton and mucus are taken by commensals. (Magnetic Island)

The pyramid shells suck body fluids from their hosts. Two *Odostomia columbianus* feeding on the mantle of *Trichotropis cancellata*. Photo by T.E. Thompson.

Above and below. *Conchodytes meleagrinae* on the gills of the pearl oyster *Pinctada margaritifera*. The close-up above shows a female on its host's gills. (2m., Tijou Reef)

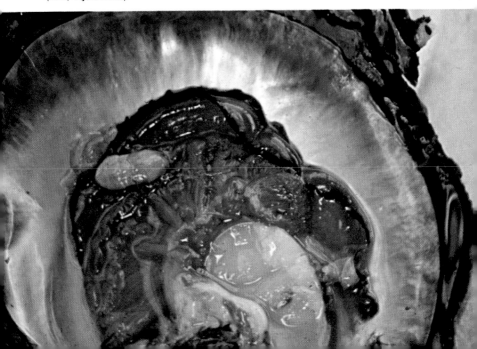

The conical shells of *Thyca*, an ectoparasite, have been found on fossil starfishes from the Devonian, and the present-day species have changed little. *Stylifer*, also ectoparasitic, lives almost completely embedded in its host's body tissues, frequently causing galls and lumps on the arms of starfishes.

Fishes

The spines of the crown-of-thorns starfish may provide shelter for a small cardinalfish, *Siphamia fuscolineata*. Although the host has been abundant, the fish has only been reported on a few rare occasions.

Small schools of these *Siphamia* live among the spines of this starfish just as other cardinalfishes live among the spines of the long-spined *Diadema* urchins. The habitats provided by both hosts are very similar and it is probable that the crown-of-thorns fish evolved from *Diadema*-inhabiting stock.

Small cardinalfish, such as *Siphamia fuscolineata,* may seek shelter among the spines of the crown-of-thorns starfish. Photo by Rodney Jonklaas. (Sri Lanka)

The messmates (*Carapus*) may also use a starfish for shelter even though they normally shelter inside sea cucumbers. *C. mourlani*, a small, elongate and almost transparent messmate, has regularly been found inside the arms of the pin-cushion star *Culcita*. *Culcita's* ambulacral grooves are narrow slits and it is difficult to force them apart without damaging the starfish, but the messmate relaxes them by simple caressing and inserts its tail little-by-little until it gets its entire body into the arm.

SYMBIONTS OF SEA URCHINS

The Echinoidea, the sea urchins, have a rounded test composed of calcareous plates and an array of sharp spines for defense. On the center of the lower surface lies a mouth equipped with a set of pincer-like teeth used for rasping sea weeds and other encrustations from rocks. Some of the tropical urchins, *Diadema* and *Echinothrix*, have very long barbed and venomous spines and are dangerous to humans. Like the crown-of-thorns starfish, these urchins attract many organisms seeking protection.

Microorganisms

Probably all echinoderms host microorganisms in their alimentary canals, but the sea urchins are especially rich in protozoans, particularly gregarines, ciliates and flagellates. Several dozen species of holotrichs are present in extremely large numbers in their intestines. Presumably many of these are parasitic, but it has been suggested that as their hosts' diets include a large amount of algae some of the symbionts may aid in the digestion of this material.

Coelenterates

A sessile ctenophore (*Coeloplana*, a relative of the ctenophore associated with soft corals) lives among the spines of the diadem sea urchins. At night when the host leaves its shelter to feed, the ctenophore creeps up one of the spines and trails its long sticky tentacles into the water, periodically retracting them to clean off the adhering plankton. *Coeloplana* obtains shelter and a mobile base which carries it into open water each night.

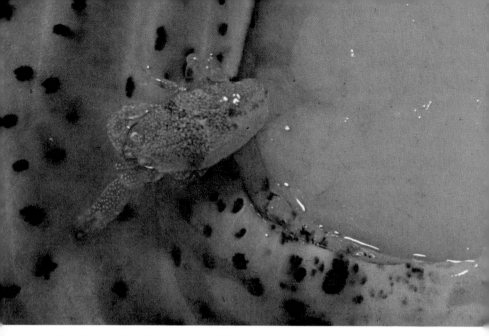

The clam shrimp *Conchodytes tridacnae* on the mantle of the giant clam *Tridacna maxima*. (3m., Tijou Reef)

Close-up of the shrimp *Anchistus* found in the winged oyster host *Magnavicula*.

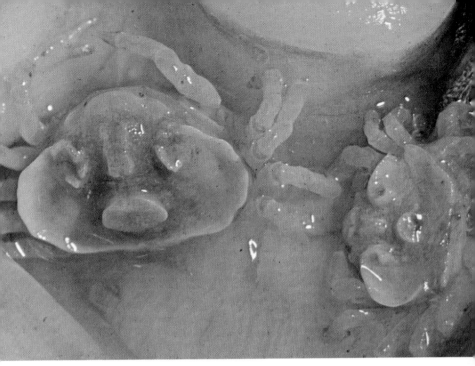

Male (smaller one) and female pinnotherid crab *Xanthasia murigera* in a *Tridacna*. (3m., Tijou Reef)

An oyster crab *Pinnotheres* laden with eggs in the tropical oyster *Crassostrea amasa*. (Intertidal, Magnetic Island)

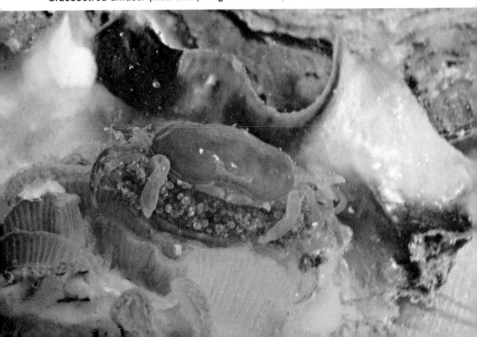

Crustaceans

Like the starfishes and most other marine invertebrates, the sea urchins are populated by many copepods. Some mobile species live on the body surfaces but others reside in spherical galls in their tests. The soft-bodied echinothurid urchins are particularly prone to infestation by these parasites.

About one dozen species of shrimps belonging to the genera *Allopontonia, Periclimenes, Tuleariocaris* and *Stegopontonia* live on the surfaces of sea urchins. Once again the long-spined *Diadema* and *Echinothrix* are popular hosts and each may support a number of species of shrimps. Shrimps such as *P. zanzibarica* and *T. zanzibarica* are both very slender and cling to their hosts' long spines, whereas the stouter *P. hirsutus* lives on the test itself.

Pairs of the beautiful red-spotted Coleman's shrimps (*P. colemani*) live on the test of the short-spined *Aerosoma*. They differ from the other urchin shrimps by their removal of their host's spines, tube feet and epithelial tissue. Theirs may be a warning coloration, for their host is venomous.

Sea urchins, particularly the long-spined types such as this *Echinothrix*, attract many organisms seeking protection. (5m., Lizard Island)

Very little is known of the biology of these shrimps. They probably feed on mucus, but their guts often contain material of similar color to their host, suggesting that they may also eat host tissues. Usually only a pair of each species will occupy each host, but large sea urchins may sustain several pairs.

Parthenopid crabs also live on certain sea urchins. Pairs of the superbly camouflaged zebra crab (*Zebrida*) are difficult to distinguish from the short-spined urchin (*Salmacis*), their host. The pair denudes areas of spines and tissues, either by the clinging action of their legs or their feeding habits. Their host, like that of Coleman's shrimps, is dangerous. These urchins have venomous pedicellaria, the small pincers which are used to remove foreign bodies from their tests. Symptoms of stinging in man are local pain, faintness, paralysis, respiratory distress and, in extreme cases, death.

A more specialized crab, *Echinoecus*, lives with the long-spined urchins. The small male lives externally near the host's anus while the larger female lives imprisoned in a gall in the host's rectum. Males eat tube feet and epithelial tissues; females eat fecal pellets and blood cells. They live in equilibrium with their hosts.

Molluscs

Several small bivalves live in association with the burrowing sand dollars. *Montacuta* lives near the anus of the spatangid urchin *Echinocardium* and probably utilizes the currents set up by the urchin.

A number of species of parasitic snails, mainly *Mucronalia*, are ectoparasites of the sea urchins. One parasite, a *Stylifer*, excavates a burrow in one of the large stout spines of a cidarid urchin and inserts its long proboscis through the test into the host's body cavity. Safely protected within the spine, it sucks body fluids from its host.

Fishes

Best known of these urchin associates is the razorfish *Aeoliscus strigatus*, which swims head down among its host's long spines. This unusual fish has a long snout, a body covered with an inflexi-

A small crab *Hypoconcha lowei* fits snugly into its protective cap, a valve from a *Pecten*. Photo by Alex Kerstitch. (18m., Morro Colorado, Mexico)

The epizoic moon mussel *Ciboticola lunata* on its host *Arca*. (2m., Magnetic Island)

The conchfish *Astrapogon stellatus* shelters in the mantle cavity of *Strombus gigas* during the day. At night it emerges to hunt for food. Photo by Charles Arneson. (Aguadilla, Puerto Rico)

A shrimp, *Anchistus,* shares its winged oyster host *Magnavicula* with a pearlfish, *Onuxodon.* (20m., Lizard Island)

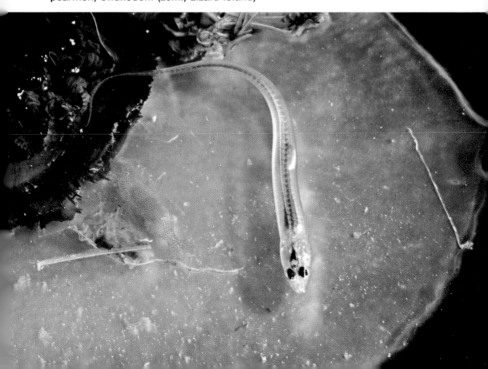

ble armor and swims mainly by paddling with its pectorals. It is laterally flattened such that when it turns toward or away from the viewer it almost disappears from sight. The razorfish's sides are striped longitudinally to resemble its host's spines.

The diadem clingfish (*Diademichthys*) has a very similar shape and habits to the razorfish although it is not related. It is a good example of convergent evolution—two organisms of different stock which occupy similar ecological niches and have evolved along the same lines. The clingfish shelters head down among the spines, is flattened (top to bottom), horizontally striped and long-snouted. Gut contents of the clingfish included the long tube feet of their host, indicating that they are predaceous.

Throughout the tropics species of cardinalfishes (Apogonidae) have entered into partnerships with the long-spined urchins. Dur-

A number of species of cardinalfish shelter among the spines of the diadem urchins. (3m., Lizard Island)

ing the daytime schools shelter among the host's spines to escape the diurnal predators, but at night when the host leaves its shelter to feed they too leave to forage. Certain species have entered a closer association with their hosts and may be seen swimming among the spines, pecking at them and at the test. Gut contents have revealed shrimp eggs and remains of the urchin's other symbionts. They may thus be performing a cleaning service for their hosts by removing commensals and parasites.

A case of an urchin eating one of its consort fish has been reported. A *Siphaluvia* was actually stabbed by a spine and transferred to the mouth and eaten. It is difficult to imagine the urchin moving a spine with the velocity needed to penetrate a fish, so perhaps the fish accidentally swam into it.

SYMBIONTS OF CRINOIDS

The delicate outstretched feathery arms of the feather stars, the Crinoidea, make them the most beautiful of the echinoderms. Those unfamiliar with them may even mistake them for flowers.

Of all the echinoderms the crinoids host the most symbionts. Their body plan is very elaborate, thus providing many nooks and crannies for fugitives. They are unpalatable to most predators and the hiding inquilines would therefore be free from indirect predation. They are ciliary-mucus feeders and capture small planktonic organisms and detritus by moving them with their mucus-covered tube feet into the ambulacral groove of the arm; there they are transported by ciliary action along open ducts to the mouth, but many commensals live along the ducts and steal this food before it reaches its destination. A few crinoids have closed ducts, possibly as a direct evolutionary response to the pirating by the commensals.

A single crinoid may support dozens of species of symbionts. One coral reef crinoid I examined had the following complement of external symbionts: about three dozen small myzostomid polychaete worms of two species; one large polynoid scale worm; a dozen small copepods; two species of shrimps; one half-crab; and three crinoid clingfish. Along with these the crinoid probably contained a rich fauna and flora in its gut, for many crinoids contain dinoflagellates and peritrich and holotrich protozoans.

Sea squirts *Herdmania momus* (center), host of the shrimp *Pontonia katoi*. (15m., Coil Reef)

The sea squirts, like the bivalves, offer symbionts protection in their branchial and atrial cavities, trapped food and mucus, and easy access via their siphons. (10m., Magnetic Island)

Pontonia sibogae on the branchial basket of the sea squirt *Styela pedata*. (10m., off Magnetic Island)

The ascidian shrimp *Pontonia katoi* in its host, the sea squirt *Herdmania momus*. (15m., Coil Reef)

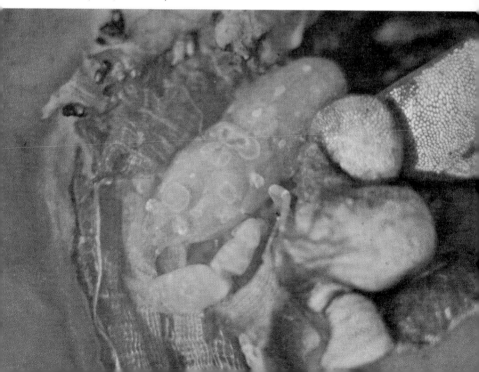

Polychaete Worms

A group of highly specialized bristle worms, the Myzostomida, occur only as commensals and parasites of the echinoderms, mainly the crinoids. They are a very ancient group, having been found on the earliest of the fossil crinoids. Myzostomids occupy a number of different niches on their hosts. Some commensal species are mobile and live on the arms and the pinnules. The parasitic species live in galls and swellings on the host's arms or in pouches on the body surface and in the gut.

The mobile species of genera such as *Myzostomum* are typically disc-shaped and fringed by sensory projections, suckers and hooks. These species generally have a long proboscis which they thrust into the ciliated grooves along which the planktonic food and mucus are transported to the mouth.

A number of polynoid scale worms, for example *Scalisetosus, Hololepidella* and *Harmothoe,* live on the arms of crinoids. These solitary, fast-moving and well camouflaged worms steal their

Positive and negative color patterns of a pair of crinoid myzostomids.

host's food and mucus but may also prey on symbiotic copepods and small myzostomid worms.

Crustaceans

Hundreds of copepods belonging to several different species may live on or in a single crinoid. Some of these, for example *Pseudoanthessius*, are not greatly modified but others, such as the worm-like *Enterognathus*, are highly specialized for a parasitic existence.

Over a dozen species of pontonid shrimps of the genera *Palaeomonella*, *Periclimenes*, *Parapontonia* and *Pontoniopsis* live on the arms of various Indo-Pacific crinoids. Usually only a pair, a male and a female, live on each host, but more may be found on larger hosts.

The shrimps are masters of disguise and may either be transparent with markings to suggest the host's arms and pinnules or be of the same rich color as their host. A single species of shrimp may be colored in bright oranges, yellows, iridescent greens, blood reds and blacks or any combination depending on the color of the particular host.

One black crinoid which had yellow pinnules contained two species of shrimps, each colored black but with yellow legs. Another host which had long thin brown arms had perfectly transparent shrimps with longitudinal brown stripes the same color and width as the arms.

Once, in deep fast-flowing water on a reef on the outer Great Barrier Reef, I found a vertical pillar of coral adorned by a single species of crinoid but of every imaginable color and hue. On the arms of each crinoid was a pair of shrimps of matching colors and hues. Or rather on all except two, for on one bright orange crinoid there was a pure white shrimp and on the neighboring pure white crinoid there was a bright orange shrimp. Clearly there had been some sort of territorial dispute before my arrival.

On the central region or oral disc of larger crinoids may live a pair of much larger pistol shrimps (*Synalpheus*) readily identifiable by their large claws. Their camouflage is less perfect, but perhaps they are too large for some of the shrimp predators or perhaps they receive better shelter. If touched, certain crinoids immediate-

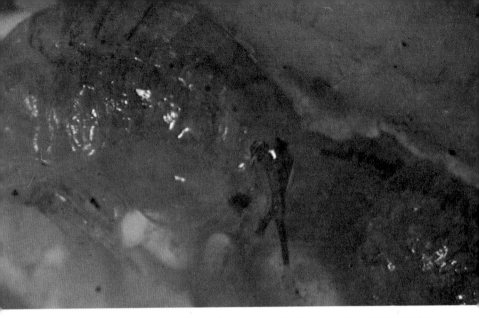

A pair of commensal amphipods in a sea squirt. Many small crustaceans live inside the ascidians. (5m., Magnetic Island)

This blue ascidian can be found attached to stalks of dead gorgonians (shown here) or other types of substrates. Photo by Dr. Patrick L. Colin. (12m., La Parguera, Puerto Rico)

Where suitable substrates are unavailable ascidians, like other sessile organisms, will settle on living surfaces. A hermit crab's shell is the substrate in this case. (15m., Lizard Island)

A dromid crab (*Dromidiopsis*) carries a cap of compound ascidians (*Atopozoa*) on its carapace. (35m., off Magnetic Island)

Synalpheus stimpsoni removed from its host. Note its massive claw which is used to produce an explosion to repel its enemies. (15m., Wheeler Reef)

ly fold their many arms over their soft central disc to protect it; the pistol shrimps may take advantage of this reaction.

The crinoid shrimps may have a variable diet. One species, *Periclimenes tenuis*, has been observed crawling over its host's arms, removing and eating attached copepods. It also feeds on mucous secretions that are rich in protozoa and bacteria. Other shrimps may steal food from the open gutters.

The half-crabs *Galathea* often live on crinoids. The beautiful *G. elegans* lives at the base of its host, usually hiding among the cirri, the claw-like appendages the host uses for anchorage. Like the shrimps, its coloration depends on that of its host. It also feeds on its host's mucus but supplements its diet with material in the surrounding waters or on the surrounding substrate. A smaller species, *G. genkai*, lives on the arms of its deep-water host.

True brachyuran crabs (*Ceratocarcinus*) live in pairs on the arms of certain crinoids. Their irregular body shapes, spines and striped coloration make them well camouflaged.

Isopods may live in and on crinoids. One of these is almost as large as the central disc but manages to crawl inside the anus to shelter there. Another lives permanently in the gut. Ascothoracid barnacles may also parasitize crinoids.

Gastropods, Brittlestars, Fishes

The crinoids may be heavily parasitized by the snails *Stylifer, Stylina, Sabinella* and *Melanella*. With their long proboscises these snails drill holes in their hosts' arms, often producing pronounced deformations, and suck out soft parts and body fluids.

Brittle stars (*Ophiomaza* and *Ophiothrix*) live at the bases of certain crinoids and drape their long arms sinuously around the arms of their hosts. They probably steal mucus and trapped plankton from the open food ducts.

The crinoid clingfish *Lepadichthys* concludes this long list of crinoid symbionts. These fish cling with their sucking discs to the upper sides of their host's arms, well camouflaged by their longitudinal stripes on a background which matches the host. They are predators, feeding on the worms, copepods and other small symbionts of the crinoid. They may also eat some host tissues, such as their pinnules.

Crinoid clingfish from a black and yellow crinoid. This host also contained a number of black and yellow shrimps and a galatheid. (25m., Lizard Island)

A coral reef eaten by the crown-of-thorns starfish. Interest was sparked in the symbiology of the starfish and other echinoderms in the hope that a biological control might be discovered. (Kangaroo Reef)

The coral-eating crown-of-thorns starfish *Acanthaster planci*, responsible for the destruction of many coral reefs in the Pacific. Photo by Rodney Jonklaas.

The starfish shrimp *Periclimenes soror* among the spines of the crown-of-thorns starfish. Photo by Roger Steene. (off Cairns)

The crown-of-thorns cardinalfish, *Siphamia fuscolineata*. When excited the dark fish assume a striped pattern. Photo by R.A. Birtles. (13m., Torres Straits)

SYMBIONTS OF SEA CUCUMBERS

The sausage-like sea cucumbers or *beches-de-mer*, class Holo-thurioidea, are superficially unlike other echinoderms but their characteristic protective spines are embedded within their fleshy body walls. They are sluggish creatures of the soft bottoms. Most species ingest sand and extract organic material, but others spread a net of tentacles in the water to trap plankton.

They are often very common on the sandy lagoons of coral reefs and have been likened to earthworms in that they are important in the turnover of sediment.

The sea cucumbers may expel long sticky threads, the Cuvierian organs, if they are disturbed. If this is not successful in deterring a predator they will disgorge their entire guts through their mouths.

The 'curryfish' sea cucumber *Stichopus variegatus* collected by a beche-de-mer diver. (4m., Palm Island)

Certain members of the genus *Stichopus* go further. If evisceration does not deter a predator they may disintegrate completely. I have held a living 'curryfish' (the pidgin name for *S. variegatus*) in my hands and after a few minutes had its flesh seemingly melt before my eyes and then drip between my fingers. This suicidal behavior is obviously disastrous for the survival of the individual but aids in the survival of the species. A predator would certainly be deterred from ever attacking another individual.

Polychaete Worms

Most of the common species of Indo-Pacific coral reef sea cucumbers are host to a polynoid scale worm, *Gastrolepida*. Generally only one of these worms lives on each host for, like other symbiotic scale worms, they are strongly territorial. They have a flattened shape and adhere to their hosts by arching their backs. However, once they leave their hosts they become efficient, sinuous swimmers.

One species of scale worm may exploit six or eight species of coral reef sea cucumbers for they assume the colors and features of their particular hosts. Those on the 'curryfish' are curry colored; those on the 'prickly greenfish' (*S. chloronotus*) are the same deep green-black of the host; those on the 'sandfish' (*Holothuria atra*) are jet black but are flecked with white spots for the host accumulates sand grains on its black body. The behavior of the scale worms which live on North American sea cucumbers was discussed earlier.

Crustaceans

A variety of crustaceans live with the sea cucumbers. The ubiquitous copepods may live in their mouths and esophagi; pea crabs may occupy their intestines; and a pair of pearl oyster shrimps may live in their cloacas.

The giant 'spiky redfish' (*Thelenota*), a sea cucumber over one meter in length and covered with many red conical projections called papillae, hosts the very beautiful king shrimp, *Periclimenes rex*. This shrimp lives among the papillae and probably feeds on the host's mucus.

The scale worm *Hololepidella* is well camouflaged among the short spines of *Acanthaster brevispinus*, a close relative of the crown-of-thorns. (35m., off Magnetic Island)

A parasitic worm in the ambulacral groove of the arm of a species of *Linckia*. Photo by Scott Johnson. (Nanakuli, Hawaiian Islands)

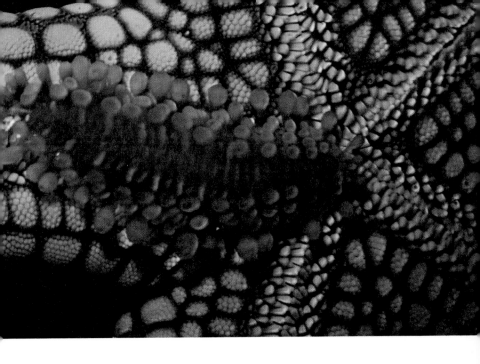

The polynoid scale worms which live on the smooth-bodied starfishes are expertly camouflaged. The parapodia of one which lives along the ambulacral grooves of *Nardoa* resemble tube feet. Another nestling between the arms of *Fromia* mimics its host's unusual coloration. (3m. Lizard Island)

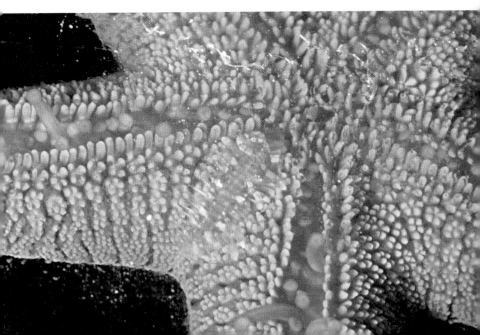

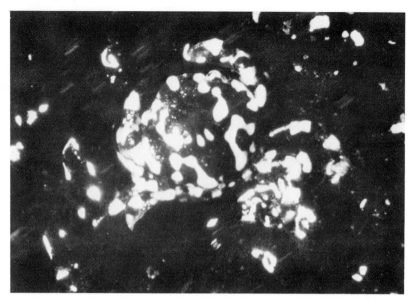

The negative crab *Lissocarcinus* on the 'blackfish' sea cucumber *Actinopyga*. The white spots resemble the sand the host coats itself with. (5m., Tijou Reef)

The emperor shrimp (*P. imperator*), a close relative of the king shrimp, may settle on a *Stichopus* sea cucumber if it cannot find its definitive host, the Spanish dancer. Similarly the pearl oyster shrimps *Conchodytes* may live in the cloacas of certain sea cucumbers if they cannot find an unoccupied pearl oyster.

A multicolored harlequin crab, *Lissocarcinus*, lives on the under surfaces or in the mouth of the 'spiky redfish.' Similar crabs live on the 'blackfish' (*Actinopyga*), but instead of being multi-colored they are black and white. Their black color blends with their host, their white spots resembling adhering sand grains. Pairs living on a host may have reverse color patterns, one the negative of the other, and they are sometimes called negative crabs.

Gastropods

The synaptid sea cucumbers are parasitized by a bivalve (*Entovalva*) and a large number of gastropods from *Mucronalia*, a little-modified genus, to *Entocolax* and *Entoconcha*, the highly specialized worm-like endoparasites. The development of the parasitism and the progressive degeneration this entails have been discussed at some length in an earlier chapter.

Fishes

The messmates (*Carapus*), once known as *Fierasfer*, are the best known associates of the sea cucumbers. They are close relatives of the pearlfish which lives within the large bivalves and are similarly elongate, scaleless and transparent.

Young fish can enter and leave through their host's cloaca head first but the larger adults must firstly insert their long tapering tails into the orifice and then wriggle and screw themselves backward and forward, gradually coaxing their way in. Once in, they swim through a pore which enters the water-filled respiratory tree or break through its wall and enter the body cavity. The holothurians possess a great capacity for regeneration, and the tear is soon repaired.

The larvae of the messmates are planktonic and swim weakly with their heads constantly pointing downward. The larval stage is called the vexillifer, so named because of a long fringed filament, the vexillum, which hangs from their necks. This filament is lost and the second stage, called the tenuis larva, begins searching for a host. Once one is found it metamorphoses into the juvenile condition. Adults locate their hosts by sensing mucus in the water currents.

Adult *Carapus* have occasionally been found with larvae of their own species in their stomachs. This cannibalism is probably a population regulatory mechanism since only one adult fish lives in each host, ensuring not too much damage is done to the host. The messmates show a progression toward parasitism. Some species supplement their diets with parts of their hosts, but the Mediterranean *C. acus* and the lipless messmate *Encheliophis* habitually feed on their hosts' gonads and reproductive tissues.

One of the pipefishes (Syngnathidae), a relative of the sea horses, has a similar way of life to *Carapus*. The messmate pipefish (*Corythoichthys*), a long slender fish about 10 cm. long, sometimes shelters in the intestines of the sea cucumbers of coastal waters of New Guinea.

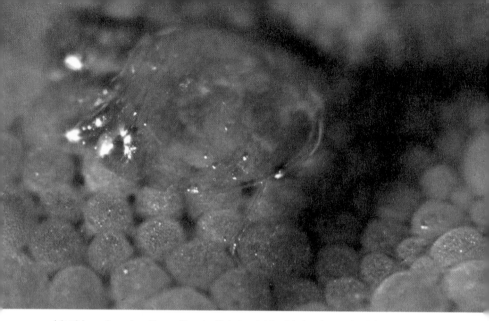

Linchiomolgus in situ on its starfish host *Linckia* is almost invisible.

Commensal polychaete worms in the ambulacral grooves of the starfish *Patiria*. Photo by T. E. Thompson. (NE Pacific)

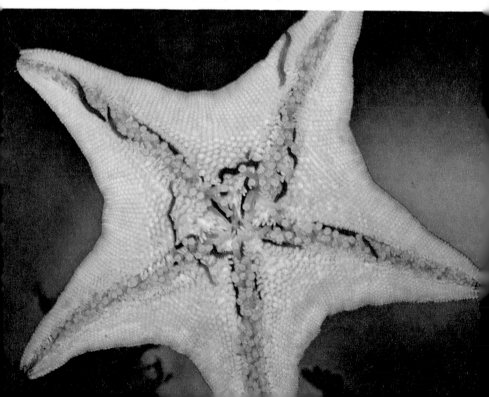

The snail *Hipponyx* attached to the arm of a starfish *Linckia multiflora*. Photo by Scott Johnson. (Nanakuli, Hawaiian Islands)

Starfish shrimp *Periclimenes noverca* entering the mouth of the short-spined starfish *Acanthaster brevispinus*. (35m., off Magnetic Island)

SYMBIONTS OF BRITTLESTARS

The brittlestars, class Ophiuroidea, resemble the starfishes but have a distinct central disc unlike the starfish. Their five arms are very mobile and are used for locomotion.

Brittlestars have previously been mentioned as symbionts of other organisms such as sponges, corals, gorgonians and crinoids, but they host fewer symbionts than the other echinoderms. This may be due to their much smaller size, their smooth body disc and lively habits.

Worms, Crustaceans, Gastropods

A number of diverse worms live on and in the brittle stars. Certain of these, for example the polynoid scale worm *Hololepidella*, are commensals and feed on their host's mucus and steal some of its food. A myzostomid polychaete, no doubt evolved from a crinoid symbiont, is also a commensal of certain European brittle stars.

The squat lobster *Galathea elegans* in its normal position between the cirri and the branches of its host. (24m., Bowl Reef)

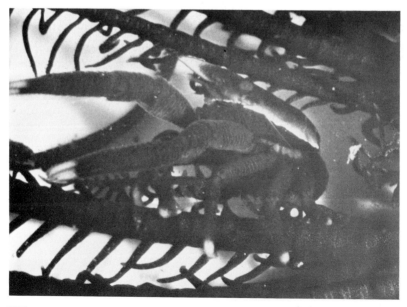

A polyclad worm is an endoparasite of ophiuroids and may castrate them by feeding on their reproductive tissues. The larvae of a number of trematode flukes use brittle stars as secondary hosts although the majority of flukes choose gastropods as intermediate hosts.

Numerous ectoparasitic asterocherid copepods live on the body disc surface of ophiuriods and more specialized species live within. Certain of these endoparasites have assumed a sac-like appearance and either can live in their host's gonads or in galls in their body cavities. Degenerate female endoparasites may be accompanied by dwarf males.

The branching-armed basket star (*Euryale*) may host the commensal shrimp *Periclimenes lanipes* among its arms. These plankton-feeding brittle stars superficially resemble the crinoids, occupy similar ecological niches and create similar microenvironments for commensals.

The gastropods *Mucronalia* and *Stylifer* which parasitize other echinoderms also parasitize the brittle stars, some of which are not much larger than themselves.

SYMBIOLOGY OF ECHINODERMS

The echinoderms host a great variety of symbionts of all types, from refugees to highly degenerate endo-parasites. The echinoderms are unusual in that whole families, even orders, of organisms live in association with them. Among these are the myzostomid polychaetes, which have evolved as crinoid commensals and parasites; the gastropod families Capulidae, Eulimidae, Entoconchidae and Paedophoropodidae; the parasitic barnacles of the Dendrogastridae; and the crabs of the family Parthenopidae.

Other groups such as the galatheid crabs, the pontoniid shrimps, polynoid bristle worms, various copepods, cardinalfishes and messmatefishes are frequently found with echinoderms but are not totally specific to them.

Why should the echinoderms attract so many diverse symbionts? There are probably many reasons. The shelter and protection afforded by the bristly sea urchins, the bushy crinoids, the sac-like sea cucumbers and certain of the starfishes attract many

A pontonid shrimp among the tube feet of a starfish. Photo by Roger Steene. (New Hebrides)

Striped starfish shrimp *Periclimenes soror* blending in with the spines of the crown-of-thorns *Acanthaster planci.* (3m., Lodestone Reef)

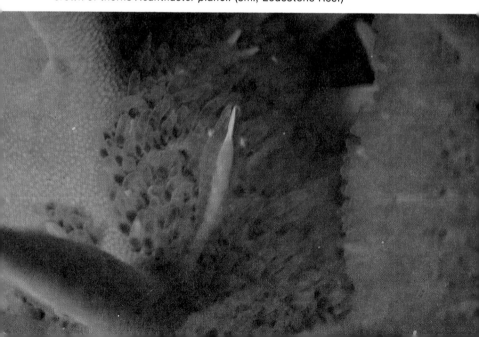

A sponge growing on the spines of an undescribed cidarid sea urchin. Other urchins have a covering of tissue on their spines to protect them from fouling. (20m., Broadhurst Reef)

Male and female parasitic snails *Mucronalia* attached to the mouth region of an echinothurid urchin. (8m., off Magnetic Island)

inquilines, particularly as the echinoderms have relatively few predators.

The mucus on the body surfaces of echinoderms is an important energy source to commensals. Like that of the corals it can be stolen without greatly harming the host. The food of certain of the echinoderms, notably the crinoids, also attracts many commensals that wish to share in it.

The echinoderms may also be able to support many commensals and parasites because of their remarkable powers of regeneration. They can rapidly repair torn tube feet, broken spines or mutilated bodies. Certain starfishes can even regenerate new individuals from fragments of arms, and the sea cucumbers can regenerate new alimentary canals and respiratory trees.

These symbionts show many and varied adaptations to their ways of life. Most of the external species are very well camouflaged, suggesting that even though their hosts may be relatively free of predators the symbionts are not. Many are strongly territorial, preventing the over-exploitation of the host's resources.

Paradoxically, although the coral-eating crown-of-thorns starfish was found to have many symbionts, including a number of parasites, no particularly destructive or pathenogenic species were ever found, thus preventing the development of a biological control. However, the most destructive phase of the plague has now passed in most parts of the Indo-Pacific; the reefs devastated (only a small proportion of the total) are rapidly regenerating, and the mysterious crown-of-thorns starfish is once again becoming a rarity.

PART IV—PHYSIO-LOGICAL MUTUALISM

Chapter 9. Symbiotic Algae

by M.A. Borowitzka Ph.D.

The seaweeds or algae form the base of the ocean's food chains. They range in size from the giant 25 m.-long kelps to microscopic unicellular phytoplankton whose size is in the range of 1×10^{-5} m. Many of these smaller algae may form symbiotic associations with marine invertebrates or other algae, and some of these symbioses are important to the biology of the host and to the ecology of the oceans in general.

Dinoflagellates are by far the most common and successful of all algal symbionts in the marine environment. This may be due in part to their thick cell wall and to the fact that they can take up and metabolize a wide range of organic compounds. Dinoflagellates can cause infections in man, largely in the eyes or the intestinal tract, illustrating their broad tolerance in environmental conditions.

The symbiotic dinoflagellate algae are commonly known as "zooxanthellae." The dinoflagellate *Gymnodinium microadriaticum* is probably the best known and most important of all algal symbionts. It is an endosymbiont of hard and soft corals and certain lamellibranch molluscs such as the giant clam *Tridacna maxima*, ciliated protozoans, anemones, gorgonians, zoanthids, nudibranchs and at least one scyphozoan (*Cassiopeia* sp.). Dinoflagellates of the genus *Amphidinium (A. chattonii, A. klebsii)* are symbionts of the hydrozoans *Velella velella* and *Porpita porpita*,

A shrimp sheltering among the spines of a venomous sea urchin. Photo by Walt Deas. (Heron Island)

Zebra crabs *Zebrida adamsi* live in pairs on a patch they wear off in the spines of the toxic urchin *Salmacis* cf. *bicolor*. Note the elaborate spines and stripes which are used for camouflage. (30m., off Magnetic Island)

A school of shrimpfish *Aeoliscus strigatus* finds shelter head down among the protective spines of the sea urchin *Diadema*. The body stripe also helps the illusion. Photo by Allan Power.

Young groupers *Epinephelus dermatolepis* also shelter among sea urchin spines. Note how their pattern of dark stripes helps hide them. Photo by Aaron Norman. (aquarium)

the turbellarian *Amphiscoleps langerhansi* and of a number of foraminifera.

Green algal symbionts are common in freshwater invertebrates such as *Hydra*, whose algal symbionts are *Chlorella* spp. In the marine environment the only green algal symbionts (zoochlorellae) are *Platymonas convolutae*, which is symbiotic with the acoel flatworm genus *Convoluta*, and a *Platymonas* sp. which, together with the dinoflagellate *G. microadriaticum*, is a symbiont of the sea anemone *Anthopleura*. Recently *Chlorella hedleyi* has been reported as a symbiont of the foraminiferan *Archais angulatus*, and it is likely that further studies will show that there are other green algal symbionts.

Blue-green algae are little-studied but common symbionts of sponges. Recently Lewin has described a new primitive alga intermediate between the green and the blue-green algae and named it *Prochloron* (Division Prochlorophyta). It is a common symbiont of tropical didemnid ascidians.

One of the most interesting and specialized symbiotic relationships is that between sacoglossan molluscs and the chloroplasts of certain larger green and red algae. In this case it is not a whole alga but only an algal organelle which is the symbiont.

One as yet little-studied symbiotic relationship occurs between the colorless dinoflagellate alga *Peridinium foliaceum* and a chrysophycean alga. This symbiosis is so complete that it at first appeared that *P. foliaceum* was a "normal" photosynthetic dinoflagellate. Detailed pigment analyses and electron microscopic studies have, however, shown that the photosynthetic component of *P. foliaceum* is an endosymbiotic alga. There are also other algal-algal as well as algal-bacterial and algal-viral symbioses, but very little is known about these associations. Further study of the organisms of our oceans is sure to yield new and fascinating examples of symbiotic relationships between plants and animals.

FEATURES OF ALGAL SYMBIOSES

It is possible to assign six common features to most of these symbioses: (1) The association is a permanent feature of the life cycles of the organisms; and (2) the associates are in intimate and sustained physical contact. These primary features permit

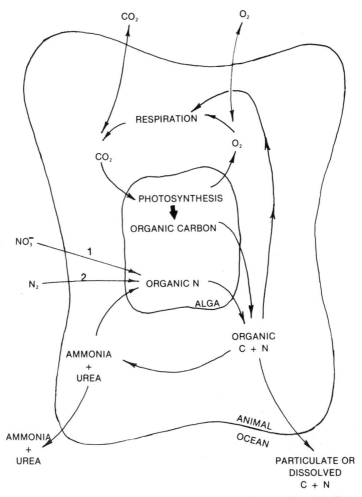

Generalized schematic diagram of the passage of nutrients and metabolites between seawater, animal and symbiotic alga or chloroplast. Note that pathway 1 does not occur in chloroplast/animal symbioses and that pathway 2 only occurs when the algal symbiont is a nitrogen fixing blue green alga.

physiological interactions such as (3) directional or two-way movement of metabolites (i.e. translocation), (4) amelioration or reduction of environmental stress, (5) morphogenetic effects and (6) appearance of synergistic properties such as the production of metabolites that are not formed by either of the organisms separately.

A myzostomid polychaete on the oral surface of a crinoid. (5m., Bowling Green Bay)

Ventral view of the crinoid polychaete *Scalisetosus longicirra* showing the long pharynx which it uses to suck food from its host's food ducts. (5m., Brewer Reef)

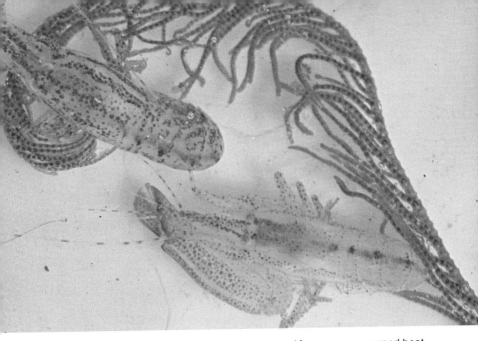

Chromatophore patterns of *Synalpheus stimpsoni* from a narrow-armed host. (25m., Lizard Island)

Pairs of pistol shrimps *Synalpheus stimpsoni* also live on the crinoids. They are larger than the other shrimps and generally live on the host's body disc. (25m., Lizard Island)

ZOOXANTHELLAE AND CORALS

The reef building corals and a large number of other invertebrates contain the endosymbiotic dinoflagellate *Gymnodinium microadriaticum*. These zooxanthellae occur primarily in the cells of the oral endoderm.

Some coelenterates such as anemones can be induced to lose their endosymbiotic algae. When these animals are then fed with tissue containing the symbionts, or when the symbionts are injected into the oral cavity, reinfection of the host occurs. It is presumed that a similar mechanism operates in corals. Once the symbiosis is established the zooxanthellae may divide and multiply. Experiments have shown that endosymbiotic zooxanthellae can divide within the host cells at a rate approaching that observed in cultures of the isolated alga. The host (coral, *Tridacna*, anemone, etc.) does not become swamped by its algal endosymbionts, as the host controls its population of algae by periodically extruding algae with its mucus so that a fairly constant number of algal cells remains within the host tissue. In cases of stress, such as decreased salinity due to rainfall on a reef flat, the coral may also extrude zooxanthellae. The host may re-ingest these extruded zooxanthellae and digest them, thereby recycling its symbiotic carbon and nitrogen.

Reef Corals: 'Super Organisms'

Reef corals are versatile in their potential ability to procure nutrients and energy as a result of their symbiosis and can therefore function at nearly every trophic level in their ecosystem: (a) the corals may behave as primary producers because the symbiotic algae photosynthesize and fix carbon at rates comparable to those of free-living algae; (b) the corals may function as primary consumers. . .since they utilize the photosynthetic products translocated from the symbiotic algae, they function to some degree as herbivores; (c) corals may also function as secondary and tertiary consumers since they ingest herbivorous and carnivorous zooplankton and bacteria as well as taking up dissolved organic matter from the seawater.

Observations of the feeding behavior of corals and calculations of their energy budget indicate that corals probably obtain no

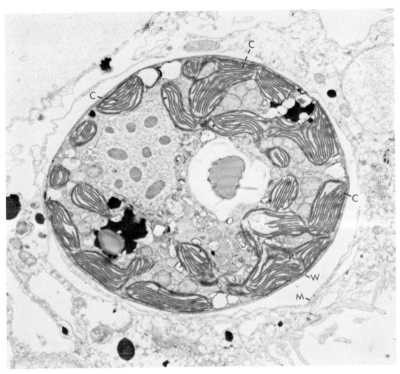

Electron micrograph of the zooxanthella *Gymnodinium microadriaticum* within the mesoglea of the staghorn coral *Acropora acuminata*. Note chloroplasts (C); algal cell wall (W); host membrane (M). M. Vesk and M.A. Borowitzka photograph.

more than 20% of their energy and carbon requirement by feeding on zooplankton. Translocation of reduced organic carbon from the phototrophic symbionts to the heterotrophic host is very likely of major energetic significance to the coral. Some corals have been shown to be able to live entirely without ingesting particulate organic material, i.e. they must be phototrophic. Calculations made by Muscatine and Porter, based on the limited experimental data available, suggest that under optimal conditions the zooxanthellae can supply 86-138% of the coral's daily carbon requirement.

Apart from the obvious importance of the algal-invertebrate symbiosis to the organic carbon supply of corals (and presumably also in other invertebrate symbioses), this symbiosis also appears to enable the organisms to survive in a nutrient-poor environment.

Crinoid shrimp *Periclimenes cornutus* from a black and yellow host. (15m., Lizard Island)

The crinoid shrimp *Periclimenes tenuis* is almost invisible when it aligns itself with its host's arm. (20m., Bowl Reef)

Periclimenes tenuis from variously colored hosts. (5m., Yankee Reef)

In certain of the symbioses with blue-green algae some fixation of nitrogen into organic nitrogen compounds occurs. In coral symbioses where the symbiotic algal partner is not a nitrogen-fixing alga the tolerance of environmental deficiencies of nitrogen is effected by adaptations for recycling nutrients within the association.

There is surprisingly little data on nutrient recycling in corals. Crossland and Barnes have shown that the algae within the coral skeletons (not the zooxanthellae, but the blue-green algae found within the calcium carbonate skeleton) fix nitrogen, and they have suggested that this fixed nitrogen may later become available to the coral. If so this would introduce a third (extracellular) partner into the coral-algal symbiosis. The zooxanthellae have been shown to be able to convert inorganic nitrogen (nitrates, nitrites) into organic nitrogen and they also excrete nitrogen-containing compounds such as alanine to the host. The uptake of host excretory nitrogen (such as ammonia) by the zooxanthellae has also been demonstrated, and there is good evidence that coelenterates containing zooxanthellae lose significantly less nitrogen to the environment than do aposymbiotic coelenterates (i.e. coelenterates without symbionts) due to this recycling between alga and animal. It is also likely that the zooxanthellae are involved in phosphate storage and metabolism.

Coral Calcification

Zooxanthellae also play an integral part in coral calcification. The classical studies of Goreau and of Kawaguti clearly showed that light stimulated calcification rate in corals. Later workers studying this phenomenon in some detail have conclusively demonstrated that this stimulation is due to the zooxanthellae and not to some other photochemical process. A number of theories have been proposed as to how the zooxanthellae are involved in calcification, but there is as yet no clear understanding of the processes involved.

The highest rates of calcification occur at the edges or at the tips of a coral colony in the region where zooxanthellae occur only in low densities or are completely absent. This high calcification rate is light stimulated and the stimulation is due to light acting on the zooxanthellae further down the coral branch. The light stimulation appears to be due to translocation of some organic compound(s) from the zooxanthellae-containing tissues to the rapidly calcifying, non-zooxanthellae-containing tissues. It seems likely that the light stimulation of calcification is due not to some direct effect of the zooxanthellae (such as a pH shift at the site of calcification brought about by carbon dioxide removal during photosynthesis) but rather that in the light the zooxanthellae synthesize specific compounds which are the energy source (e.g. ATP) or a calcium-containing compound or matrix material or something similar.

PLATYMONAS AND *CONVOLUTA*

The association between the green alga *Platymonas* and the flatworm *Convoluta* is one of the best studied algal symbioses. The green prasinophycean alga *Platymonas convolutae*, when cultured

The green alga *Platymonas* is cultured in the tissues of the flatworm *Convoluta*. The worms migrate out of sand during the day so their zoochlorellae might photosynthesize. (Intertidal, Townsville)

The parasitic isopod *Cirolana lineata* entering its host's anus. (10m., Keeper Reef)

The isopod *Cirolana lineata* removed from its host's hindgut. Note a young isopod (below) which has escaped from the female's brood pouch.

Galathea elegans crawling on the arms of a crinoid. Photo by Walt Deas. (Heron Island)

Close-up of the same species, *Galathea elegans* at the base of a crinoid. (Keeper Reef)

independently of its host, has four flagellae and a thick theca (cell wall). Upon ingestion by the host several significant morphological changes occur. The alga loses its eyespot, flagellae, flagellar pits and theca, producing a naked protoplast. The loss of the theca permits a very close association between the alga and the host tissues. This means that solutes produced by the alga need only pass across a single membrane from alga to host. Experiments have shown that the alga releases compounds such as glucose, mannitol and lactic acid, as well as some amino acids, to the host. This symbiosis allows *Convoluta* to exist for most of its life without apparently feeding. Young *Convoluta* do not have algal symbionts; these must be ingested anew at each generation. Once established the symbiosis persists until the death of the animal.

Green algae other than *Platymonas convolutae* may form symbiotic relationships with *Convoluta* in the laboratory, but none of these "artificial" symbioses are nearly as effective as the "normal" symbiosis. This suggests that the host cells cannot specifically recognize the "normal" algal symbiont. However, it does indicate that there is something about the "normal" symbiotic alga's genetic makeup which enables it to form a better and more efficient symbiosis with *Convoluta*. In the field only *Platymonas convolutae* has been observed as a symbiont of *Convoluta*.

CHLOROPLASTS AND SACOGLOSSAN MOLLUSCS

Sacoglossan molluscs such as *Elysia* and *Placobranchus* feed by piercing the algal cell with a single radular tooth and then sucking out the algal cell contents including cell organelles such as nuclei, mitochondria and chloroplasts. The fact that these molluscs are mainly found on the large-celled coenocytic algae *Codium*, *Bryopsis*, *Caulerpa* (Chlorophyta) and *Griffithsia* (Rhodophyta) is probably a direct result of their mode of feeding: algae with smaller cells require a proportionately greater effort to obtain the necessary nutrients.

Once ingested the cell contents are moved by ciliary action to the digestive gland where they are taken by phagocytosis into the digestive cells to be digested. However, in many sacoglossans, especially the elysids, the chloroplasts are not digested. Instead the

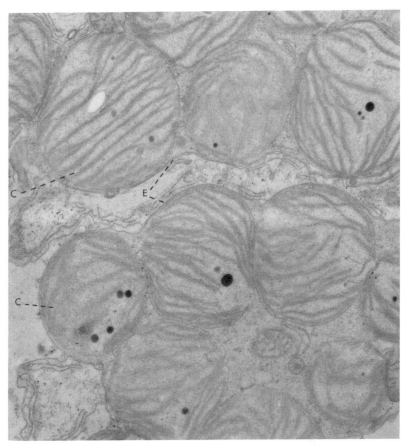

Symbiotic chloroplasts within the sarcoglossan *Elysia maoria* found feeding on the green siphonaceous alga *Codium*. Note chloroplast (C); chloroplast envelope membrane (E). R. Hinde photograph.

chloroplasts pass further into the host cell and continue to photosynthesize, remaining within the host cell for several months. It is not known whether the chloroplasts are able to divide within the host cell; as they do not synthesize the green photosynthetic pigment chlorophyll they probably do not divide.

When the sacoglossan and its endosymbiotic chloroplasts are exposed to the light the plastids photosynthesize, fix carbon dioxide and evolve oxygen. Up to 50% of the carbon fixed by the chloroplasts may be transferred to the host where it is metabolized

Alpheid shrimp commensal with a crinoid. Photo by Walt Deas.

Commensal brittle star *Ophiomaza cacaotica* lives under crinoids and steals food from their open food ducts. (35m., off Magnetic Island)

Crinoid clingfish *Lepadichthys lineatus* and associated crinoid. The clingfish seems at home at the base (above) as well as out along one of the arms (below). Photo above by Bruce Carlson. (Fiji); photo below by Dr. V.G. Springer (Gulf of Aqaba, Red Sea)

and incorporated into mucus, protein, etc. The degree of importance of these symbiotic chloroplasts to the energy budget of a sacoglossan mollusc has as yet not been determined although it appears that in some molluscs with symbiotic chloroplasts growth is retarded in the absence of light. One very fascinating and as yet unanswered question is the problem of why some sacoglossans digest their chloroplasts and others maintain them. How do the digestive cells recognize an algal chloroplast as distinct from mitochondria or nuclei? Can they differentiate between the plastids from different algae?

CONCLUSION

What does the alga get from the animal? The obvious answers are that the host supplies the algal symbiont with a relatively protected and stable environment. The host also supplies carbon dioxide and nutrients for the alga's metabolism while the alga in turn supplies the host with reduced carbon compounds, organic nitrogen compounds and oxygen.

Endosymbiotic relationships are often seen as a model for the possible evolution of eucaryotic cells and their membrane-bound organelles such as chloroplasts and mitochondria. Chloroplasts are proposed to have evolved from a symbiotic association between a non-photosynthetic cell and a procaryotic alga. This symbiosis eventually became so specialized that the endosymbiotic alga became a permanent feature of the "animal," thereby turning it into a "plant"! The symbiosis between the colorless dinoflagellate *P. foliaceum* and its chrysophycean endosymbiont has some features which one would expect that the ancestral animal/plant must have had. Most modern symbiotic relationships are, however, between two eucaryotic organisms, but much may still be learned of how symbioses between two quite unrelated (taxonomically) organisms become established and how they function.

PART V.
VERTEBRATE SYM-
BIOSIS
Chapter 10. Anemonefishes and Anemones

Anyone who gazes into a rockpool at the sea shore will almost certainly see borders of delicately colored flowers along the sides and in crevices. If one drops a pebble onto a crown of petals the flower immediately comes to life, closing on the foreign body only to reject it soon after as inedible. These flowers are of course the animals known as sea anemones.

Like other cnidarians (coelenterates excluding the ctenophores) the sea anemones carry batteries of microscopic stinging cells or nematoblasts, already discussed. A small fish or crustacean brushing against the innocuous-looking tentacles will trigger batteries of nematoblasts which inject a powerful venom into the victim. The dead or paralyzed prey is then carried by the tentacles to the central mouth and swallowed. Inside the anemone's simple body cavity, the coelenteron, the food is digested. Any wastes are expelled through the mouth.

ANEMONEFISHES: THE CLOWNFISHES

It is very strange to see beautifully colored and seemingly defenseless little clownfishes (*Amphiprion*) luxuriating among the deadly tentacles of the giant tropical sea anemones of the Indo-Pacific.

Juvenile cardinalfish sheltering in a large crinoid. (30m., off Cape Bowling Green)

Ventral view of the crinoid clingfish *Lepadichthys caritus* swimming along the arms of its host. Its pelvic sucker is clearly visible. (5m., Yankee Reef)

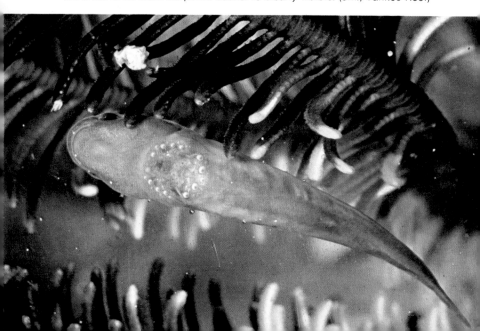

Biologist collecting symbionts from the sea cucumber *Thelenota ananas*. Many new associations were discovered while researching this book. (3m., Wheeler Reef)

The sea cucumber scale worm *Gastrolepidella clavigera* on its host *Stichopus variegatus*. (4m., Palm Island)

These unmistakable fishes, striped with almost luminous blue-white on an orange or red background, parade back and forth above the tentacles of their anemone host. If another fish or a human observer comes too close the clownfish dart down into the canopy of tentacles and then peer out cheekily from their refuge, assured that nothing would pursue them.

The symbiosis between clownfishes and giant anemones was first described by C. Collingwood in 1868. Much has been written about the partnership since then, mainly about the mysterious way the little fish avoids being stung by the deadly tentacles. The most comprehensive text on the clownfishes is that by G. Allen.

Biology of Clownfishes
TAXONOMY

The clownfishes are placed with the damselfishes (Pomacentridae) although some workers prefer to accord them full family status (Amphiprionidae). There are 27 described species, all very similar except for the sabre-toothed clownfish, which is usually placed in its own genus, *Premnas*.

The clownfishes have an Indo-West Pacific distribution and are thought to have evolved in the Indo-Australian-Philippine zone where the largest number of species, ten, is found. From here they are thought to have radiated from reef to reef in stepping-stone fashion. Remote reefs were never reached, probably because the duration of the larval stage is not very long.

Clownfishes have formed partnerships with 13 species of anemone belonging to the genera *Cryptodendrum, Parasicyonis, Physobranchia, Radianthus,* and *Stoichactis*. These actinians, reaching a meter or more in diameter, have varying morphologies, distributions, habitat preferences, diets and nematoblast potencies. Likewise the different species of clownfishes have different distributions, host preferences, sizes, feeding habits and behavior.

The narrow-striped clownfish, *A. perideraion*, is the smallest species. It is host specific, cryptically colored to blend in with the tentacles and rarely leaves their security. On the other end of the scale Clark's clownfish, *A. clarki*, is much larger, boldly colored, lives in a variety of host species and frequently leaves the protective mantle of the tentacles.

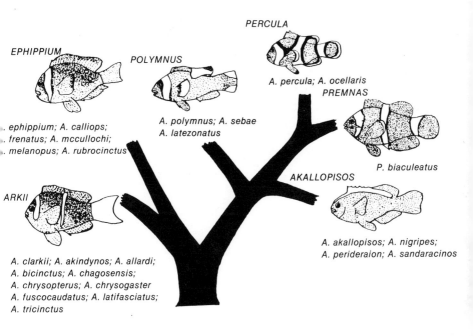

EPHIPPIUM

POLYMNUS

PERCULA

A. percula; A. ocellaris

PREMNAS

. ephippium; A. calliops;
. frenatus; A. mccullochi;
. melanopus; A. rubrocinctus

A. polymnus; A. sebae
A. latezonatus

P. biaculeatus

ARKII

AKALLOPISOS

A. akallopisos; A. nigripes;
A. perideraion; A. sandaracinos

A. clarkii; A. akindynos; A. allardi;
A. bicinctus; A. chagosensis;
A. chrysopterus; A. chrysogaster
A. fuscocaudatus; A. latifasciatus;
A. tricinctus

The proposed evolution of the species of clownfish. (Redrawn from Allen)

BEHAVIOR AND LIFE HISTORY

The clownfishes have strong pair bonds and, once established, have a life-long association with their host. Usually a single male and female adult pair of a species occupies a host, but if it is a large anemone a number of smaller individuals are also present. More than one species may be found on a single host.

When entering a reproductive condition a pair interrupts foraging to engage in courtship displays of postures, rolls, approaches and fin displays. The female elicits nest-building behavior by nudging the smaller male, who then begins to clear an area near the anemone.

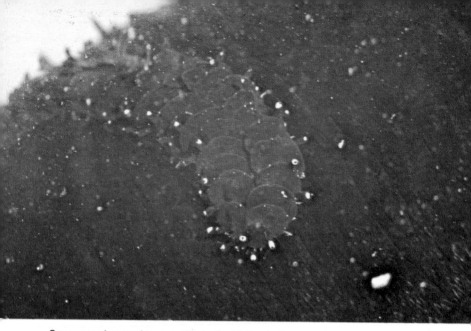

Sea cucumber scale worm *Gastrolepidella clavigera* on *Holothuria atra*. (3m., Lodestone Reef)

Detail of the head of *Gastrolepidella*.

The scale worm *Gastrolepidella clavigera* matches the coloration of a number of sea cucumbers. This host is the large *Thelenota ananas*. (7m., Lodestone Reef)

Undescribed commensal ophiuroid on the surface of the 'prickly red fish' *Thelenota ananas*. (7m., Lodestone Reef)

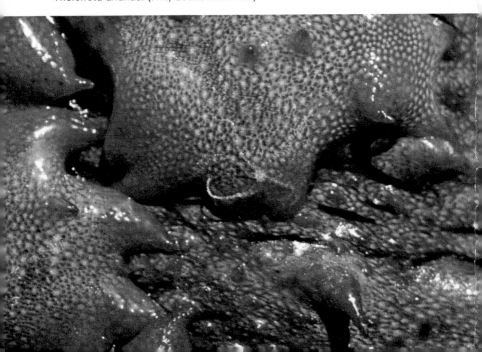

The female lays 200-300 eggs and the male swims over them to fertilize them. The eggs are tended by the male, who removes the dead ones and circulates water around them. He mouths them, possibly to coat them with mucus so they will not be stung by the host (see 'Immunity' below). He is very aggressive at this stage, driving off intruders with rushes and a clicking noise. One night, after about a week, the eggs hatch and the small planktonic young drift away.

The larvae spend about two weeks in the plankton and then begin a search for a host using visual and olfactory stimuli. If they find a host they have to acclimate to acquire the protective mucus. If the host a larva adopts is crowded by other clownfish it is driven off, and if it fails to find another host it is probably soon eaten. However, if there is room it joins the other clownfish at the lowest rung of the peck order.

Recently Allen has found that many of the 'juvenile' clownfish living with mature adults are, in fact, adults themselves. The established pair which dominates them somehow suppresses their growth so that the subordinates become midgets. However, if the dominant individuals die then the midgets once more begin growing. This is an ingenious mechanism of preventing the overexploitation of the host while maintaining a potential breeding pool for use in contingencies.

Biology of the Partnership

BENEFITS TO CLOWNFISHES: PROTECTION, FOOD, CLEANING

The clownfishes obviously benefit from the protection offered by the anemone's stinging cells. Their host would kill or seriously injure any predator daring to pursue them into the tentacles. Clownfishes have never been recorded living in the absence of a sea anemone—the partnership for them is obligatory.

Clownfishes removed from their host and placed well away from it soon fall victim to the coral reef piscivores. Likewise, clownfishes placed in a mixed aquarium will also be eaten by one of the other fishes. However if the clownfish and the anemone partner are placed in a similar aquarium then they will be safe from predation.

The distinctive colors of most of the clownfishes advertise the fact that they are thus protected. These aposematic colors signal potential predators not to come too near for they have a dangerous partner. The colors are bright so the clownfish will be easily recognized and not mistaken for a vulnerable fish.

Clownfishes also receive other advantages from the association. The anemone host may provide them with food. At least some of the clownfishes nibble at tentacles, eat mucus, sloughed epithelial tissue, cell fragments and waste products, including food scraps. Studies of gut contents reveal nematoblasts and zooxanthellae, materials of host origin. However, the clownfishes more often feed on planktonic material which drifts past or on detritus from the surrounding bottom.

The clownfishes may also benefit by having their parasites stung to death by their anemone. Allen, however, noted that the clownfishes were not free from parasites. Indeed, they seemed to have as many as any other damselfish. Isolated clownfishes are

The giant Indo-Pacific sea anemones, some over one meter in diameter, host small clownfishes which live with immunity among their deadly tentacles. (15m., Bowden Reef)

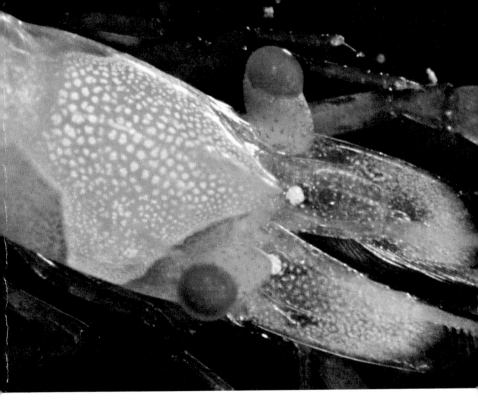

The beautiful king shrimp *Periclimenes rex* lives among the spines of *Thelenota ananas*. (5m., Wheeler Reef)

The harlequin crab *Lissocarcinus orbicularis* on the tube feet on the underside of *T. ananas*. (4m., Yonge Reef)

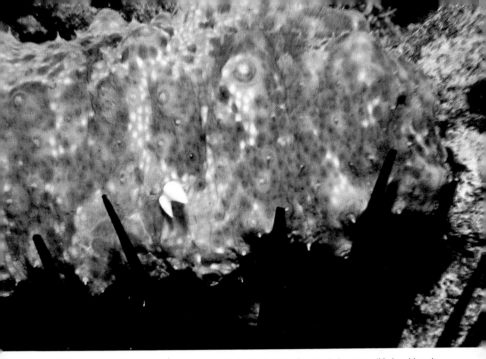

Balcis parasitizing a sea cucumber. Photo by Scott Johnson. (Koko Head, Hawaiian Islands)

The pearlfish *Carapus bermudensis* and its associated host *Astichopus multifidus*. The pearlfish comes out at night to feed but shelters within its host by day. Photo by Dr. Walter A. Starck II. (Florida)

very prone to diseases in the aquarium, but it has been suggested that this occurs for 'psychological' reasons. In the absence of the host the clownfishes often fret and feel insecure and may therefore have a low resistance to diseases.

The clownfishes apparently take great pleasure from luxuriating among the anemone's tentacles and 'bathe' in them for many hours at a time. A student of clownfish behavior, R.N. Mariscal, considers that the pleasure of contact with the host is a 'tactile reward system' to ensure a continuous contact with the mucus on the anemone's tentacles. The reason for this behavior is described below.

BENEFITS TO ANEMONE

Although the clownfishes need the anemone for their survival, the anemone does not need the fish, for many healthy fishless anemones may be seen around the coral reef. Nevertheless it is thought that the host does benefit in some ways.

The main benefit is that it is cleaned by its partner, which, as described above, eats sloughed and necrotic tissues, mucus and food wastes. The clownfishes may also eat the anemone's parasites, and their bathing behavior may flush away silt and aid in water circulation.

The last way the clownfishes may help their host is a matter of some conjecture. It is commonly held that the fish feeds its host in return for the protection it receives. Clownfishes in the aquarium have often been seen doing this—I have seen it myself. However, Allen doubts that it occurs very often in nature. Despite many hundreds of hours studying clownfish behavior in the Pacific, he has never seen an instance of this behavior. On the other hand Mariscal, who studied Indian Ocean species, has seen host-feeding in the fish he has investigated.

THE PARTNERSHIP: SUMMARY

The clownfish/sea anemone partnership is a multifaceted one with elements of shelter and protection (inquilinism), food sharing (commensalism) and cleaning (mutualism). The conflicting evidence of the details of the symbiosis could easily stem from differences in the behavior of the 27 species of fishes and 13 species of host.

IMMUNITY TO NEMATOBLASTS

The ability of the clownfishes to safely live among the tentacles of the anemones when other fish are stung to death has puzzled biologists for a century.

Many theories have been proposed to explain this immunity. Some held that the power lay with the fish; others held that it lay with the anemone. It has been suggested that the anemone 'learns' the identity of the fish from behavioral, physical or chemical stimuli and suppresses the discharge of the nematoblasts. The main objection to this theory is that the coelenterates simply do not possess the neural equipment for such behavior.

Supporters of the alternate theory suggested that the fish might avoid direct pressure on the tentacles or might carry antigens to give it immunity to the stings. Some skeptics even doubted that the nematoblasts were powerful enough to affect the fish.

The black anemonefish *Amphiprion melanopus* among the tentacles of its host anemone. Photo by Roger Steene. (Michaelmas Reef)

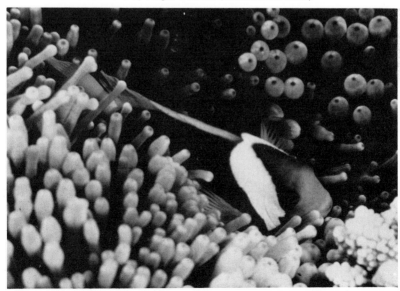

A small ectoparasitic *Stylifer* attached to a spine on the arm of the brittle star *Ophiarachna incrassata.* (Torres Strait)

A large parasitic *Mucronalia* on the brittle star *Macophiothrix.* (Torres Strait)

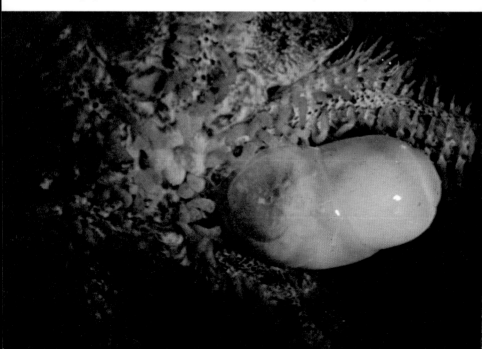

Coral reefs result from a plant/animal partnership. (Ribbon Reefs)

A partnership between animal and plant has resulted in the evolution of reef-forming corals and giant clams. The dinoflagellate *Gymnodinium microadriaticum* is an endosymbiont of giant *Tridacna* clams and hermatypic corals. (2m., Lizard Island)

Most of these ideas were advanced by scientists who had never even seen a clownfish. A series of simple and logical experiments over the last decade provided the solution: the active member is the clownfish, which coats itself with a layer of the anemone's mucus to become "anemone-like," for the host does not sting itself.

This was found by first placing chunks of freshly killed clownfish among the tentacles. If placed flesh-side down the nematoblasts discharged, but if placed skin-side down they did not. Clearly something in the skin provided the protection.

In another experiment mucus from a clownfish was removed by a sponge and the sponge was thrust among the tentacles. A detailed examination revealed no discharged nematoblasts on it. The de-mucused clownfish was reluctant to make contact with the tentacles and was stung when it was forced into them. Clearly the mucus of the clownfish held the key.

Clownfishes were then isolated from their hosts for varying periods and later returned to them. It was found that fish isolated for less than an hour were not stung but those isolated for longer were stung. Fish isolated for 20 hours had lost all immunity. The protective factor in the mucus therefore depended on contact with the host.

The fish which had been isolated had to get to "know" their host all over again. They did this by first making the briefest of contacts, usually with their tails, but by their sudden recoil it was obvious that they were stung. They then made a slightly longer contact; although they were stung again it was not as severe. Thus, little-by-little the duration and intensity of the contacts increased until they could once more luxuriate among the tentacles. The process of gradual change was termed "acclimation."

Anemonefishes therefore either physiologically alter the mucus on their scales or coat themselves with some factor from the host. The latter has proved to be the case. Hosts have been labelled with radioactive isotopes and a transfer of the isotope to the fish has been detected.

It is now known that the anemonefishes coat themselves with the host's mucus and become "anemone-like." Some factor in the host's mucus inhibits the discharge of nematoblasts, otherwise they would be constantly firing into other tentacles whenever they

314

made contact. The nematoblasts must identify other parts of the anemone, and this has been exploited by the clownfish.

OTHER ANEMONEFISHES

The clownfishes are the best known and most specialized of the associates of anemones, but a number of other fishes also live among their tentacles.

Dominofishes

Schools of juvenile dominofish, *Dascyllus trimaculatus*, a damselfish of the Indo-Pacific, shelter very close to and sometimes among the tentacles of the Indo-Pacific giant anemones. These

Subadult dominofish leave the protection of the anemone but may remain in the vicinity. (Lodestone Reef)

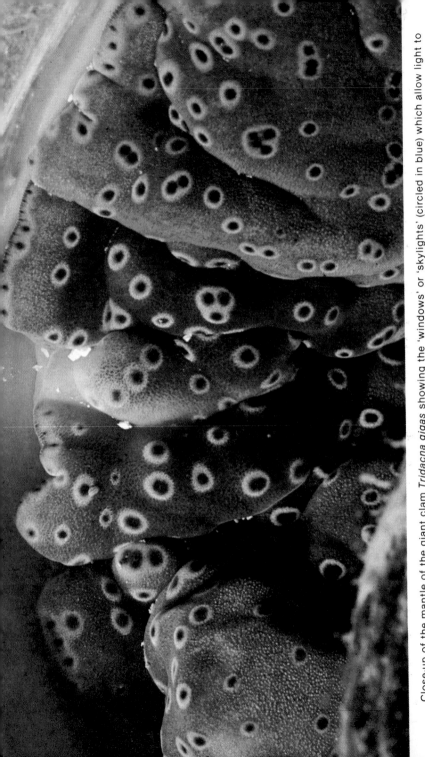

Close-up of the mantle of the giant clam *Tridacna gigas* showing the 'windows' or 'skylights' (circled in blue) which allow light to penetrate to the 'greenhouses' within. Photo by Dr. Gerald R. Allen. (Euston Reef)

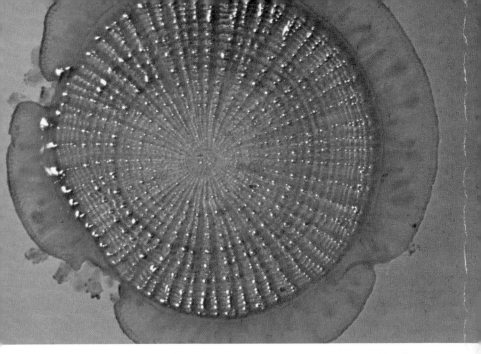

The round-rafted *Porpita porpita* (above) and the by-the-wind sailor *Velella velella* (below) harbor the dinoflagellate *Amphidinium.* (Surface, Evans Head, New South Wales)

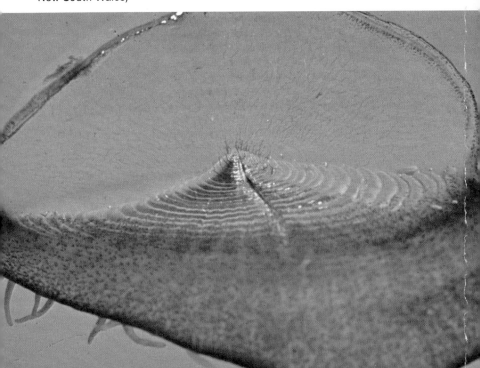

fish, jet black with white spots, are also very conspicuous and if threatened dart into the tentacles. Presumably they too possess the 'host-recognition' mucus although they do not bathe in the tentacles for long periods. As the school grows older it spends less and less time among the tentacles. The adults do not hide in the tentacles but are often found in the vicinity of the anemone. More often they shelter in the branching coral *Pocillopora*.

Blennies and Gobies

Some of the small bottom-dwelling blennies and gobies have taken advantage of the protection offered by sea anemones. The small Mediterranean goby *Gobius bucchichii* hides among the tentacles of the snakelocks anemone, *Anemonia sulcata*, and is also thought to coat itself with its host's mucus. In the Caribbean several species of the blenny genus *Malacoctenus* shelter in the giant anemones, and there are reports of similar fishes sheltering in anemones which host Great Barrier Reef clownfishes.

Other Fishes

A number of coral reef fishes have regular although less intimate contacts with giant anemones. Some occasionally make contact with the tentacles and may even acquire the anemone's mucus, but most fishes carefully avoid the tentacles, relying more on protection through proximity. At least twenty species of reef fishes, mainly cardinals (*Apogon*), wrasses (*Thalassoma*) and butterflyfishes (*Chaetodon*), have been reported to have a loose association with great anemones.

THE ANEMONE COMMUNITY

In addition to the clownfishes and other fishes, the giant anemones host many invertebrates in their tentacles.

The giant Caribbean anemones harbor blennies among their tentacles, and wrasses and cardinals live in close proximity. At their bases live pistol shrimps and crabs: Pederson's and Yucatan cleaner shrimps live on tentacles and wave their long white antennae at passing fish to attract them.

The giant Indo-Pacific anemones have an even more impressive collection of symbionts. A study of a community in sub-tropical

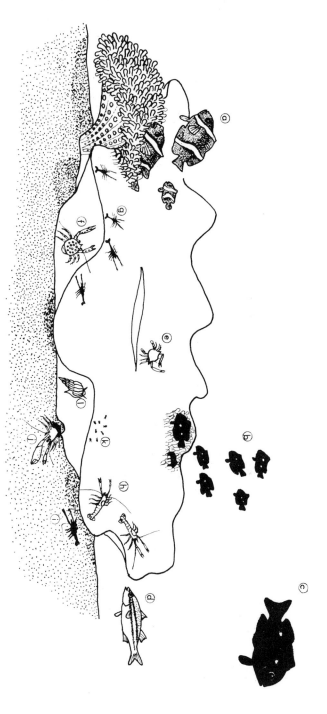

The symbiont community of the giant sea anemone *Stoichactis haddoni*. (a) Clownfish *Amphiprion akindynos*. (b) Juvenile dominofish *Dascyllus trimaculatus*. (c) Adult dominofish. (d) Cardinalfish *Apogon*. (e) Portunid crab *Lissocarcinus laevis*. (f) Half crab *Petrolisthes maculatus*. (g) Shrimp *Periclimenes holthuisi*. (h) Shrimp *P. brevicarpalis*. (i) Shrimp *P. nilandensis*. (j) Pistol shrimp *Alpheus*. (k) Copepod *Linchiomolgus myoroe*. (l) Stair case shell *Epitonium*.

319

Sea anemones, such as this *Calliactis* which is attached to pumice stone, culture *Gymnodinium* in their tissues. (Surface, Magnetic Island)

Blue-green algae invade the calcareous tests and skeletons of marine invertebrates such as these sand dollars. (Intertidal, Magnetic Island)

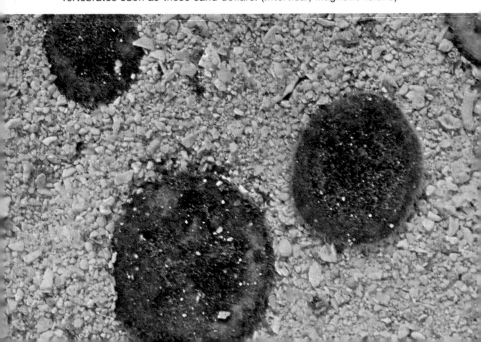

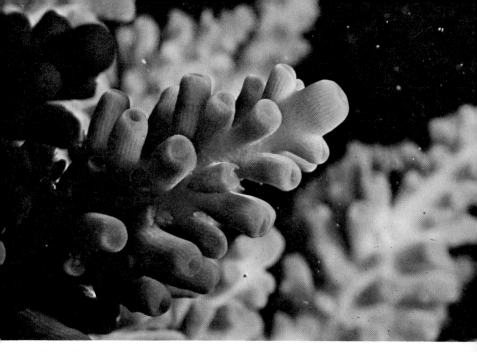

The highest calcification rates occur at the tips of the branching coral *Acropora*. (3m., Lizard Reef)

Under the stress of low salinity corals such as these *Acropora* may expel their colored zooxanthellae. (4m., Magnetic Island)

Moreton Bay, Australia revealed that the anemone *Stoichactis haddoni* had, in addition to the usual clownfishes and dominofishes, the following: seven species of true shrimps, a mysid shrimp, a porcelain crab, a true crab and several species of endozoic algae. Organisms living in a looser association include: five species of fishes, six species of crabs and three species of shrimps. In all, a grand total of thirty or more obligatory and facultative symbionts! I have also seen copepod crustaceans and parasitic wentletraps (*Epitonium*) on this anemone.

Like the symbionts which exploit the crowded branching corals and gorgonians, the anemone symbionts live in different parts of the host (on the tentacles, among them, under the fleshy lobe holding the tentacles, at the base of the anemone, in the water near it). Likewise they take different things from the host (shelter, mucus, sloughed tissue, zooxanthellae, food wastes, other symbionts, body tissue, body fluids). Many probably also restrict their own numbers (but some are gregarious) and other factors such as predation by the anemone cleaners must act to keep their numbers down.

Anemones and Crabs

The smaller sea anemones of the tropical and temperate seas also appear in a number of symbioses, most commonly with the hermit crabs.

Throughout the world's seas members of the hermit crab superfamily Paguroidea (mainly *Pagurus* and *Dardanus*) have formed alliances with members of the sea anemone family Hormathiidae (mainly *Calliactis*, *Paracalliactis* and *Adamsia*). These anemones are found on the snail shells worn by the hermit crabs to protect their vulnerable soft abdomens or, in special cases, on the abdomens themselves.

The hermit crabs gain protection from the anemone's nematoblasts as well as camouflage. The sea anemones on the other hand gain from the mobility that the crabs give them: they are brought into contact with more food as they ride along on the hermit crabs and sometimes take a share in the crab's own meal. Hermit crabs are scavengers that shred their food. The anemones on the shell above often lean over to take some fragments.

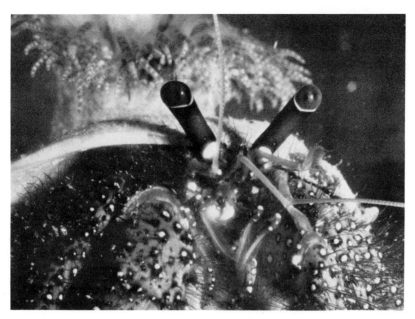

The sea anemone *Calliactis miriam* protects the hermit crab *Dardanus* and receives food from the crab. (Intertidal, Lizard Island)

In some cases the association may be initiated by the crab, in other cases by the anemone. The anemone *Calliactis parasitica* detaches itself from its hold on a rock when the hermit crab *Pagurus bernhardus* moves past. The agile anemone somersaults onto the shell and adheres to it. Most of the other crabs actively seek out anemones and place them on their shells. Anemones usually cannot be removed without damaging them, but the wily crabs relax them by stroking their columns before carefully picking the anemones off the rocks and putting them on their shells.

Hermit crabs must keep changing their shell size as they grow. When the time comes for a change they seek out a suitable unoccupied shell and quickly swap them. Those with symbiotic anemones stroke their partners to relax them and then plant them on their new shells.

Shell swapping is always a dangerous time for hermit crabs, but *Pagurus prideauxi* has solved the problem. This crab does not have to swap shells because as it grows its sea anemone partner, the beautiful *Adamsia palliata*, secretes a horny material to enlarge the

Like terrestrial flowers these soft corals (8 pinnate tentacles) and hard corals (6 or multiples of 6 tentacles) reach for the sun. Their zooxanthellae make them primary producers of the reef. (3m., Lizard Island)

Tomato clownfish *Premnas biaculeatus.* Photo by Roger Steene. (Michael-mas Reef)

Amphiprion akindynos above a large *Radianthus.* The small fish are prevented from maturing by the established adults. (20m., Bowl Reef)

shell. This anemone is large and effectively protects the entire abdomen of its host. The crab, while feeding, has been seen to place fragments of food into its partner's tentacles. The association is mutualistic and neither partner can live without the other. The symbiosis may also operate at a physiological level as it has been shown that the hermit crab has a natural resistance to the anemone's toxins.

A very similar association to the anemone/hermit crab one, and one possibly ancestral to it, is that between anemones and snails. In European waters *Calliactis parasitica* may settle on a whelk shell occupied by a living snail rather than a shell occupied by a hermit crab. It has been shown that the anemone finds the shell by locating a chemical produced in the periostracum or outer layer of the shell. This substance is an insoluble protein; it is thought that the anemone does not use a chemoreceptor but instead recognizes it by "feeling" its molecular configuration.

In Greenland waters a third partner has joined the snail and anemone. A nemertean worm lives under the pedal disc which the anemone uses for attachment to the snail's shell. The worm is immune to the host's nematoblasts and may even enter the anemone's mouth to steal food from its gastral cavity.

EVOLUTION

The anemone/hermit crab partnership may have evolved from the anemone/snail partnership. An ancestral anemone which lived on a snail may have mistakenly settled on a snail shell occupied by a hermit crab. The latter, benefiting from the association, might later have taken an active role.

It is more likely that the symbiosis evolved with the hermit crabs, which may have decorated their shells for camouflage and protection. At some later date an anemone which settled on a hermit crab's shell may have mistakenly settled on a living shell.

Chapter 11.
Pilotfishes and Suckerfishes —
Escorts of Sharks

Pilotfishes and suckerfishes, the escorts of the large open-water fishes and occasionally of marine reptiles and mammals, have been objects of curiosity and sometimes veneration since those days when man first put to sea in fragile fishing craft. The two types of fishes seem very different in morphology and behavior but they are, in fact, very much alike. Both are hitchhikers, commensals and probably cleaners.

PILOTFISH

The boldly striped pilotfish acquired its name from its habit of stationing itself in front of large predatory fish, seemingly to lead them about. Popular folklore has it that the pilotfish guides sharks to their prey and is given scraps of food and protection from its enemies in reward.

"The great grey robber-shark, his black fin hoist,
Like pirates sail, and slimy belly of pearl;
A spear blade gleaming as it cuts the blue,
The little fishes fly save one bold sort
Striped motley, with long snout, which is the slave
And lick-plate of the shark, seeking for him
Food, that the little fish may leavings eat;
No shark so hungry that will swallow him."

(Edwin Arnold: *The Voyage of Ithobal*)

The ancients revered them, believing that the fish similarly guided or piloted their vessels during sea passages. The pilotfish

Amphiprion perideraion in its host anemone *Radianthus ritteri*. Above: the anemone is almost closed leaving the fish somewhat exposed to danger. (New Guinea) Below: A large anemone with an entire "family" of *A. perideraion*. Enewetak, Marshall Islands). Photos by Dr. Gerald R. Allen.

Anemones may at times be shared by different species of fishes. Here the dominofish *Dascyllus trimaculatus* may be seen sheltering near an anemone that also houses several specimens of clownfish *Amphiprion perideraion*. Photo by Allan Power.

joined their boats shortly after they put to sea and left them shortly before they arrived at their destinations. The pilotfish is, in fact,
an open-water species.

Greek poets told of the power of pilotfish as long ago as 350
B.C., and the Roman historian Pliny wrote of its habits at about
the time of Christ. It is featured in the legends of oceanic peoples;
for example, the Torres Strait Islanders named a constellation
after it—*Wapi*, the two bright stars which lie in front of *Baidam*,
the great shark constellation.

Pilotfish participated in the famous *Kon-Tiki* expedition. During the drift of the balsa raft from South America across the
Pacific, the adventurers often captured large sharks and the or-
phaned pilotfish adopted the raft. At one point 60 or 70 sheltered
under the raft, and some were thought to have accompanied it
4,000 sea miles. The escorts swam in formation beneath their
adopted host and ate kitchen scraps thrown overboard. The rafters
regarded them affectionately:

> "These queer little fish huddled under our protecting wings
> with such childlike confidence that we, like the shark, had a
> fatherly protective feeling towards them. They became the
> *Kon-Tiki's* marine pets, and it was tabu on board to lay hand
> on them."

Biology of Pilotfish

There is probably only one species of pilotfish, *Naucrates* ('ruler
of ships') *ductor* ('leader'). It has a world-wide distribution but is
more common in the tropical oceanic waters. Its distribution is
dependent on the presence of suitable hosts, mainly oceanic
sharks.

The pilotfish is a jack or trevally (Carangidae) similar in shape
and size to most others of its family. Its color is singularly distinctive for it has about eight dark cross-bands on silvery sides and a
blue dorsal surface. It grows to about 60 cm. in length and is a
palatable food fish.

Little is known of the life history of the pilotfish. The young are
marked similarly to the adults but have a spiny operculum quite
different from that of the adults. The juveniles were not recognized as pilotfish for many years and were placed in a different

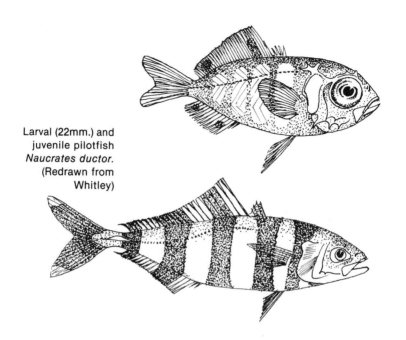

Larval (22mm.) and juvenile pilotfish *Naucrates ductor*. (Redrawn from Whitley)

genus. Juveniles of a few centimeters in length shelter under drifting flotsam and jellyfishes, especially the dangerous Portuguese man-o-war (*Physalia*). They have been seen making contact with the tentacles and even nibbling them, suggesting that they are immune to their host's nematoblasts. At a later stage they leave their first hosts and adopt large predatory fishes, their hosts during their adult life.

Reports on their reproduction and early life history are conflicting. It has been stated that the eggs have a long tail and are actually laid on the host, but another report contends that the eggs are round, are heavily laden with oil, and float freely in the plankton.

The adults are territorial, their territories being their mobile hosts. The largest adults are dominant over the others associated with the host, and when two hosts near each other and territories overlap there are squabbles among the clouds of pilotfish and some are driven from their hosts. This intra-specific competition occurs only when two or more hosts are near enough so that the vanquished fish has the opportunity of joining another host.

The fish change color during these disputes. During the threat display the characteristic dark bars fade, leaving the fish silver below and dark blue above.

Unlike *Amphiprion akindynos*, the orange clownfish *A. ocellaris* rarely leaves the safety of its host's tentacles. (Lodestone Reef)

Anemonefishes make their nests on the substrate near their host. Here a male *Amphiprion perideraion* hovers above its eggs. Photo by Dr. Gerald R. Allen.

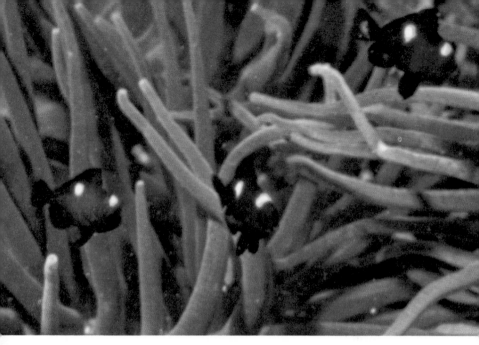

The young of the dominofish *Dascyllus trimaculatus* shelter near, and occasionally in, the giant sea anemones. Photo above by Dr. Fujio Yasuda; photo below by the author. (15m., Lodestone Reef)

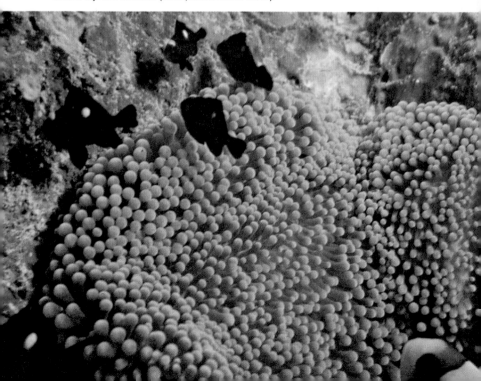

Benefits To Pilotfish

TRANSPORT

The traditional notion of the pilotfish leading their hosts to prey is now in disrepute. Although they station themselves in front of their hosts and appear to lead them, they do this to ride on the pressure wave set up by the powerful host. They hitchhike on the boundary layer just as dolphins hitchhike in front of ships. In this way they can cover great distances in search of food without expending much energy.

FOOD

Stomach contents of pilotfish indicate that they are opportunistic feeders. They mainly eat small fishes which they actively capture. Because they travel great distances with their hosts they are probably brought in contact with more food than a similar free-living fish of the open sea. Pilotfish also eat scraps of food from the shark's meal. Sharks often tear their prey into pieces and fragments drift around, freely available to the commensal pilotfish. Pilotfish also eat their host's feces.

PROTECTION

The pilotfish undoubtedly gain protection by living in close proximity to a savage predator. What fish would dare snatch a pilotfish from the very jaws of a shark? The conspicuous coloration, like that of the clownfishes, acts as a warning to its predators: try to eat me and you will be sorry! This aposematic coloration is common in the animal kingdom and serves to warn potential predators that the wearer is dangerous, poisonous or noxious in some way.

Benefits to Host

Iconoclastic modern science has established that the pilotfish is merely a hitchhiker which eats its host's food scraps and even its feces. It is no longer a 'ruler of ship' or 'leader.' But does it contribute anything to its host?

Most biologists are of the opinion that the pilotfish contribute little or nothing to the partnership, but I am of the opinion that they do. The evidence, largely circumstantial, is based on the following reasons:

(1) There is some evidence that pilotfish clean their hosts. They are opportunistic feeders and may take parasites as well as decaying food scraps.

(2) The pilotfish are rarely, if ever, eaten by their host, suggesting that it is in the shark's interest not to eat their escorts. Something which over-rides the legendary feeding instincts of sharks must be very important.

(3) The vivid coloration of the pilotfish may be an identifying signal to the shark as well as a warning to potential predators.

Whitetip oceanic shark *(Carcharhinus longimanus)* and a school of pilotfish. Foreground: pilotfish with transient coloration acquired during intra-specific disputes. (Drawn from various sources)

The dominofish *Dascyllus trimaculatus* will often leave the protection of the anemone for "cleaning" by the cleaner wrasse *Labroides dimidiatus*. Photo by Dr. Gerald R. Allen. (Palau Islands)

Although sea anemones kill and eat fishes, certain fishes and crustaceans can live among their tentacles. (Magnetic Island)

The transparent shrimp *Periclimenes brevicarpalis* has colored disruptive blotches. Male and female pairs live on the giant anemones of the Indo-Pacific. Photo by Roger Steene. (Pixie Reef)

P. brevicarpalis on an adopted host in an aquarium. (Intertidal, Magnetic Island)

It has been shown that cleaning symbiosis is very important to fishes and that cleaners are immune to predation. The coloration of the pilotfish may be a 'guild sign' identifying it as a cleaner to the predatory shark. However, biologists are divided on the matter of the immunity to predation of pilotfish. Some consider that the pilotfish are not eaten because they are too fast for their host, but these scientists could never have witnessed the lightning speed of a feeding shark.

One of the most experienced diving biologists, Walter Starck II, believes that the distinctive barred coloration of the pilotfish and sea snakes is mainly a disruptive one rather than an identifying one. Starck believes that the broad high-contrast bands running at right angles to the axis of motion confuse the sharks. Sharks have an 'image intensifier' in their eyes which greatly increases contrast, enabling them to see farther underwater. Because of this they might only be able to see the dark bars and mistake them for small individuals swimming in formation.

The barred coloration has also been adopted by a relative of the pilotfish, the juvenile Indo-Pacific golden trevally, *Gnathanodon speciosus*.

Golden Pilotfish

Postlarval golden trevallies are tightly schooling fish and may shelter near drifting flotsam, large fish or even a diver. As they grow they become more solitary but still continue to swim with large fishes of the coral reefs and surrounding waters.

Like *Naucrates*, juvenile *Gnathanodon* often station themselves in front of their hosts to ride on the pressure waves. The resemblance is so close that the majority of divers, including biologists, mistake them for true pilotfish. Their hosts include the giant cod and groupers, reef sharks, whale sharks, manta rays, other large fishes and some mammals such as dugongs.

When it reaches about 15-20 cm. in length the golden trevally ceases to pilot and loses its definitive bars and yellow-gold background. The adults, faintly barred on a pale yellow background, are schooling fish of reef waters.

On a number of occasions I have been fortunate enough to witness the piloting behavior of the golden trevally. Once, while I

Golden trevallies before becoming adult school near drifting flotsam, large fishes, or any other object living or dead that would offer them shelter. Photo by G. Marcuse.

was examining a crown-of-thorns starfish, a school of these fish came from around the corner of a coral outcrop and inspected me. I attempted to photograph them but my threatening movement startled them and they retreated around the corner from where they had come. I followed closely behind them only to see them fleeing into the gloom accompanied by a giant Queensland grouper. I am quite sure that the grouper never actually saw me or sensed threatening vibrations, yet it was warned of my presence by the flight behavior of its golden pilots.

It is possible that the pilotfish similarly warn their hosts of the presence of other large fishes, either sources of danger or prey. A school of pilotfish often surrounds the host like a wing of escort fighterplanes screening a bomber. Surely six or twelve extra sets of sensory receptors are better than one. It is feasible that the host might respond to changes in the pilotfishes' behavior and be alerted to danger or food.

The small shrimp *Thor amboinensis* can be found on or around various sea anemones which offer them protection. Photo by Dr. Patrick L. Colin. (3m., Discovery Bay, Jamaica)

The anemone mysid shrimp *Heteromysis actinae* swarming around the tentacles of *Bartholomea annulata* along with the alpheid shrimp *Alpheus armatus* (with red and white banded antennae). Photo by Dr. Patrick L. Colin. (10m., Eleuthera Island, Bahamas)

Periclimenes yucatanicus, one of the cleaner shrimps, is seen here associated with *Bartholomea annulata* although it may also be found with other anemones. Photo by Dr. Patrick L. Colin. (4m., Discovery Bay, Jamaica)

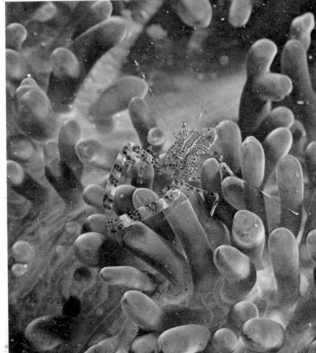

Another anemone shrimp *Periclimenes* sp. on the anemone *Stoichactis helianthus.* Photo by Dr. Patrick L. Colin. (12m., Aguadilla, Puerto Rico)

Evolution of Pilotfish

The strong urge of many of the fishes of the open waters to seek shelter near some objects has been discussed. The open sea is almost completely devoid of shelter. It is characterized by its homogeneity, its 'sameness', and the many predatory fishes seeking a meal. Thus anything drifting, a coconut, a piece of pumice stone or detached seaweed, is used by a variety of species for shelter.

The young of the trevallies have a definite habit of sheltering near flotsam. Some species, slightly more specialized, associate with jellyfishes. Other species shelter near fishes from time to time and may actually 'pilot' the moving host, taking advantage of the pressure wave.

The vertical bars of the pilotfish are a development of the typical faint stripes of most of the juvenile trevallies. The piloting behavior is likewise a development of the trend for juvenile trevallies to associate with living and non-living objects drifting in the open sea. I suspect that the pilotfish has evolved neotenously, its larval or juvenile characteristics having been retained in the adult fish. Not only are they retained, but they are highly refined.

SUCKERFISHES

The suckerfishes or remoras (family Echeneidae) occupy a similar niche to the pilotfish although few fish could be more different. Whereas the pilotfish is streamlined, graceful and startlingly patterned, the suckerfishes are long and slender, weak but sinuous swimmers, and are usually drably colored. While the pilotfish ostensibly 'lead' their hosts, the suckerfishes clamp onto them with their suckers to obtain a free ride.

They too have been objects of interest for thousands of years but unlike the pilotfish they have usually been treated with scorn. The ancient Greeks believed that the suckerfishes exerted a mysterious power over their vessels and could slow them down or even stop them. This belief persisted with the Romans, who accused them of causing the downfall of Antony by slowing down his boat during the battle of Actium. Pliny described how they stopped the Emperor Caligula's galley and how sails and 400 slaves could not budge it. In more recent times they were also accused of stopping sailing boats, even on windy days. (Boats lose steerage at times or mysteriously stop, but this is due to a layer of 'dead' freshwater covering a layer of heavier salt water moving in the opposite direction.)

The suckerfish's power to delay things extended beyond the water. They were used for love potions and aphrodisiacs, and pregnant women in some cultures stay away from fish markets for the sight of a suckerfish could delay the birth.

However, in some places around the world the suckerfish were treated with reverence as they were used to catch fish and turtles. In the Caribbean, the Torres Straits, and in other islands of the Pacific suckerfishes were captured, a line was tied to their tails, and they were released again to find a host. When they clamped onto a host the fisherman pulled the pair in, ensuring that he kept tension on the suckerfish at all times. Suckerfishes which performed well were kept as pets in special pools or in flooded canoes, and their owners often spoke to them and praised them, hoping that they would continue to work well.

A juvenile *Echeneis naucrates*. It is a weak swimmer, has a filamentous tail and is a uniform gray. It will rapidly mature into an adult. (Magnetic Island)

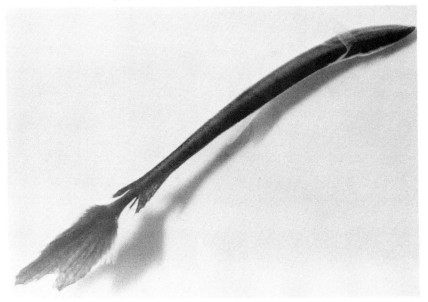

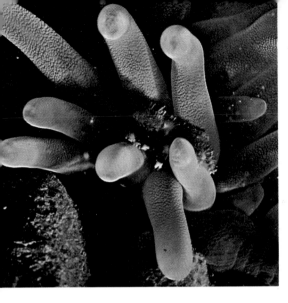

The anemone crab *Mithrax cinctimanus* is immune to the stings of the sea anemone *Condylactis gigantea* as it hides within the tentacles. Photo by Dr. Patrick L. Colin. (2m., Discovery Bay, Jamaica)

This spotted half-crab *Petrolisthes maculatus* lives on the trunk at the base of the giant anemone. (2m., Lodestone Reef)

The porcelain crab *Petrolisthes* sp. living among the tentacles of one of the larger sea anemones of the genus *Stoichactis*. Photo by U. Erich Friese.

P. maculatus also lives among its tentacles. Photo by Roger Steene. (New Hebrides)

Biology of Suckerfishes

The suckerfishes belong to the family Echeneidae, a very ancient group of fishes. They are elongate with minute scales and have a protruding lower jaw and a unique oval sucker on the top of their heads. The sucker is actually a highly modified dorsal fin which migrates and reforms during the larval development of the fish. The fin rays form lamellae or slats rather like those of a venetian blind. When the slats are raised they create a vacuum in the disc and this clamps the fish onto its host. Forces of attachment of 10-15 kg. have been measured. Little energy is expended in creating this force for the drag imposed on the hitchhiking fish forces it backward, thereby lifting up the lamellae and increasing the suction—the faster the host swims the greater the suction. The fishermen therefore kept a force on the captive fish to increase the suction.

Although about 60 species have been described it is probable that only about ten are valid. Indo-Pacific species include: the slender suckerfish, *Echeneis naucrates*, which attaches to sharks and large fishes only when it is young and exists free-living when adult; the short suckerfish, *Remora remora*, which lives on sharks and marlin; the lousefish, *Phtheirichthys lineatus*, which lives on barracuda; and the pale suckerfish, *Remoropsis pallidus*, which spends its adult life living in the gills of marlin where it eats parasites and scraps of its food.

The suckerfishes lay small eggs and the larvae are free-living in the plankton. When the larvae are one or two centimeters in length the suckers begin to form and they begin to search for a suitable host. Some attach to flotsam for an interim period, perhaps waiting for a suitable host to present itself.

Biology of Suckerfishes/Host Symbiosis

The nature of the symbiosis is not fully understood. The weak-swimming suckerfishes clearly benefit by their hitchhiking for they gain transport. Like the pilotfish, they also gain protection because of their proximity to an impressive host.

The suckerfishes are also opportunistic feeders taking small crustaceans and fishes from the water, scraps of food from the feeding host and parasites which grow on their host. Once, when I

346

was photographing a giant grouper and golden trevally pilotfish, a flash bulb exploded, spreading glass fragments about me. From the underside of the grouper a small slender suckerfish detached itself and came over to peck at the glittering glass fragments, no doubt thinking they were scraps of fish scales.

Cleaning Behavior

There is strong evidence to suggest that the suckerfishes are important cleaners of the large open-water fishes. The dentition and the superior position of the mouth and its long lower jaw suggest that these are adaptations for removing parasites while remaining attached and parallel to the host. Suckerfishes have been seen entering the mouths and gills of their hosts, and gut contents have revealed caligoid copepods, parasites of fishes. They are also largely immune from predation, a good indication that they are necessary for the survival of their hosts in some way.

Hans Hass reported that one of his companions had the unpleasant experience of having a suckerfish try to bite off his nipple, no doubt because it resembled an ectoparasite.

A marine copepod from the skin of a tiger shark. Such parasites may be removed and eaten by the pilotfish and suckerfish. (Knuckle Reef)

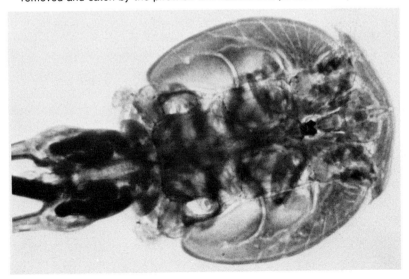

347

The beautiful boxer crab *Lybia* uses captured sea anemones to gather food and to sting predators. Photo by Bruce Carlson. (Fiji)

Two anemones *Adamsia palliata* attached to the shell housing the crab *Pagurus prideauxi*. The crab gets additional protection, the anemones get mobility. Photo by U. Erich Friese.

Young pilotfish, *Naucrates ductor*, showing the distinctive bars of the pilot guild. Photo by Dr. Shih-chieh Shen. (Taiwan)

Juvenile carangids piloting a giant whale shark *Rhincodon typus*. Photo by J. Veron. (Ribbon Reefs)

My own experiences with suckerfishes add support to the theory that they are important cleaners. I once obtained two very young specimens of slender suckerfish, *Echeneis naucrates*, only about 4 cm. long, attached to a pair of needlefish (*Tylosurus*). The suckerfish were very thin, slate gray in color and had long veil-like tails. They thrived in an aquarium, where they adopted other fishes, often to their annoyance.

The young fish were initially very poor swimmers but as they grew older they became more mobile and would slide all over their hosts, nibbling and nudging them. I would hand feed them (they ate finely diced shrimps and fish) and once they became used to me they would groom my hands, nibbling at hairs and loose skin. The grooming tickled and I could well imagine the pleasure or annoyance it might give a willing or unwilling host.

About a month after their capture (by this time they had doubled in size) they underwent a sudden metamorphosis. One day they were gray with long tattered tails, but the following day when I fed them they were handsomely striped with a longitudinal black stripe lined top and bottom with white and their tails were quite rounded. I was amazed at the transformation, but what struck me was their uncanny resemblance to the striped coral reef cleanerfish. Surely the possession of the common cleaner guild sign indicates that they are cleaners.

Evolution of Suckerfishes

The affinities of the suckerfishes are unknown. It has been suggested by some that they evolved from the cobia, *Rachycentron*, which they resemble. The echeneids might have evolved from weak-swimming stock which sheltered and rested under drifting flotsam, perhaps resting their heads on it to conserve energy. The dorsal fin could have gradually evolved into a device for holding the fish on the floating object. With this sucker the suckerfish could also rest on moving fishes to be transported along. The opportunistic feeders might then have exploited the host's ectoparasites and a mutualistic association could have begun. The peak of this specialization is represented by the pale remora, which could be well on the way to becoming a degenerate cleaner and/or parasite of other fishes.

Chapter 12.
Cleaners—Key Animals In The Fish Community

Of all the fishes on coral reefs, and possibly in other communities as well, the most important are no longer than a man's finger. These fishes, the cleanerfishes, remove and eat ectoparasites, diseased and injured tissue and unwanted food scraps from those fishes which visit them. They occupy the special niche of 'fish physician' and enjoy a unique status in the community—they are partly or wholly immune from predation. A glimpse of the special colors of the cleanerfishes will override the predatory urge of the hungriest grouper or moray eel.

The inordinate importance of the cleaners was graphically demonstrated to me some years ago on a coral reef of the Great Barrier Reef. The sky was overcast and the roaring southeast trade winds heaped up a gray, choppy sea. I was skin diving near the reef edge and had been lured farther and farther out to sea by a school of batfish. The water deepened and the sandy bottom fell from sight, leaving me surrounded by a diffuse green envelope of water. Unnerved I began the long swim back to the reef edge when suddenly in front of me a vast shape detached itself from the gloom. For a few seconds wild thoughts flashed through my brain. . .only a giant tiger shark or a great white could be that big. And then the apparition became clearer—it was a huge manta ray three or four meters in wing span, ghosting through the water with lazy sweeps of its huge fins.

It was oblivious to me and, awed, I followed at a safe distance as it headed toward the reef edge until a prominent head or 'bommie'

351

Juvenile *Gnathanodon* and *Echeneis* sheltering near a resting shark *Ginglymostoma*. Photo by Dr. Dwayne Reed. (Fiji)

The highly conspicuous coral reef pilotfish, the golden trevally *Gnathanodon speciosus*. The host is a 2 meter long Queensland grouper *Promicrops lanceolatus*. Photo by R.A. Birtles. (20m., Bowl Reef)

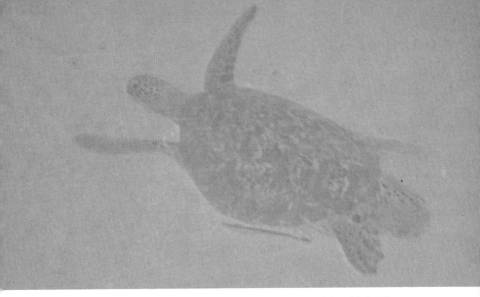

Suckerfish hitching a ride on a hawksbill turtle. (15m., Lizard Island)

This juvenile *Echeneis naucrates* has assumed the longitudinal stripes of the adult and has begun cleaning other reef fishes (Magnetic Island)

of coral half as large as a house came into view. To my amazement the giant stopped in mid-water two meters from the coral face, opened its cavernous mouth and spiralled its "devil's horns" which are used to direct its planktonic food into its mouth. It seemed to enter a state of torpor.

From a number of places on the coral head a dozen or more small blue and black striped fish converged on the manta. They swarmed all over it, entered its mouth and disappeared among its gills. I could now get close enough to look into its mouth and I dared to touch its hard, rough fins, but the ray took no notice. Biologists avoid teleological descriptions—interpreting animal behavior in terms of human desires and emotions—but I am sure that the manta was in a state of ecstasy.

The manta remained suspended for ten minutes and then it came to life again, quivered and, as one, the little fish (blue-streak cleaners, *Labroides dimidiatus*) left it to return to the coral head. The manta itself then left, disappearing into the green gloom.

Blue streak cleaner wrasses *Labroides dimidiatus* grooming *Manta alfredi*. The suckerfish under the giant may also be cleaners. (5m., Heron Island)

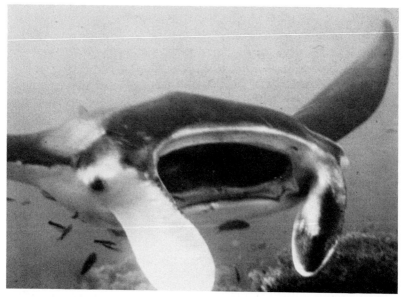

One of the giants of the sea, *Manta alfredi,* visits a cleaner station to have its huge mouth cleaned. (5m., Heron Island)

I became a regular visitor to the coral head and several times each day I saw the manta, sometimes accompanied by two or three smaller ones, visit the cleanerfish. I found myself musing: *How important these little fish must be if a giant like this manta, a fish weighing as much as a hundred thousand times their weight, should revolve its life about them.*

Although I have seen many instances of cleaning behavior since then, the sight of that ecstatic giant remains vivid in my mind.

DISCOVERY OF CLEANING SYMBIOSIS

The discovery of cleaning symbiosis in the sea and its great importance is a recent and exciting one.

Cleaning behavior among terrestrial animals had been observed more than two thousand years ago by the first naturalists, the Greeks. Herodotus wrote of a bird which actually entered the open mouth of the Nile crocodile to remove and eat parasitic

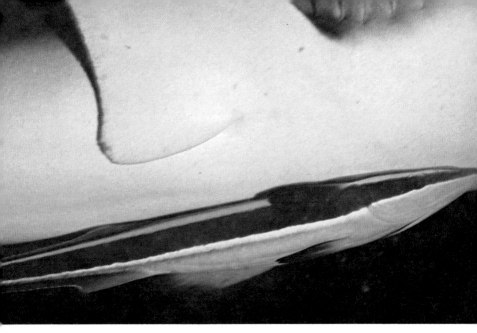

The slender suckerfish *Echeneis naucrates* hitchhiking on a black-tip reef shark *Eulamia spallanzini*. (Magnetic Island)

White-tip reef shark *Triaenodon apicalis* with escorting trevally and sucker-fish. (5m., Wheeler Reef)

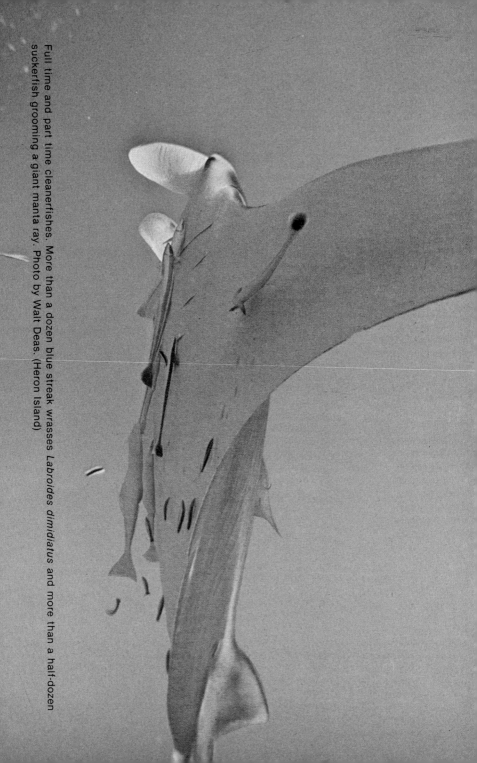

Full time and part time cleanerfishes. More than a dozen blue streak wrasses *Labroides dimidiatus* and more than a half-dozen suckerfish grooming a giant manta ray. Photo by Walt Deas. (Heron Island)

leeches. They also knew of the cattle egrets which would stand on the backs of grazing cattle and pick off ticks. In recent times fishermen called certain small Mediterranean and Atlantic fish barberfish because they groomed other fish.

The adventurer/scientist William Beebe, famous for his pioneering descents into the ocean deeps in a bathysphere, observed in 1924 that red shore crabs in the Galapagos Islands would regularly remove and eat ticks living on the sun-bathing iguanas. Four years later and half a world away in the Caribbean he saw essentially the same behavior among reef fishes: small wrasses of certain species would clean the bodies and algae-covered teeth of large parrotfishes which sought them out.

Other cases were reported in the 1930's. An attractive little fish, the blue-streaked *Labroides dimidiatus* from the Indo-Pacific, was seen cleaning the other inmates of the Amsterdam aquarium. Skindivers in the Caribbean reported that the small black-and-blue-striped gobies (*Gobiosoma (Elacatinus)*), juvenile wrasses (*Thalassoma*) and porkfish (*Anisotremus*) also cleaned reef fishes.

Labroides dimidiatus grooming the gills of this cooperative batfish *Platax*. Photo by Muller-Schmida.

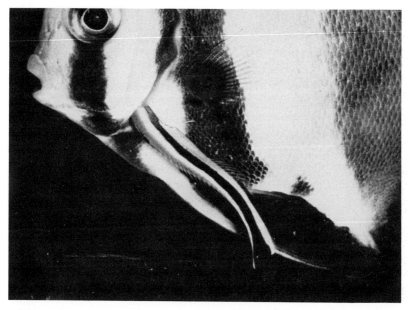

The advent of SCUBA in the early 1950's finally permitted man to observe fishes in their natural surroundings. One of the pioneering diver/photographers, Conrad Limbaugh, was the first to recognize the great importance of cleaning symbiosis. Limbaugh saw flocks of sea gulls alight on the backs of basking sunfish in Monterey Bay to pick off their parasites, and beneath the surface of the bay he saw little damsels, señoritas, remove and eat the parasites and clean the wounds of large predatory pelagic fishes which visited them in the kelp beds.

Limbaugh discovered that cleaning was even more common in the warm waters of the Caribbean coral reefs. He identified about a dozen species of butterflyfishes (Chaetodontidae), gobies (Gobiidae) and damsels (Pomacentridae) as cleaners during either their juvenile stages or throughout their entire postlarval lives. Their hosts included most of the fish in the coral reef community.

Suspecting the importance of cleaning, Limbaugh removed all known cleaners from a small isolated coral reef. The results astonished him. Within a few days the numbers of fishes had declined markedly and those which remained were badly infected with fungal growths, ulcers and had frayed fins. In a complementary experiment he added fish cleaners to an aquarium in which the fishes were in poor health. Within days their condition had improved.

Although Limbaugh's early work and certain of his generalizations have been criticized by the more objective scientists who succeeded him, his discovery of the great importance of cleaning symbiosis in the sea ranks as one of the most important recent discoveries in marine ecology.

IMPORTANCE OF CLEANING

Cleaners are now recognized as key animals in the coral reef community because almost every other reef fish is dependent on them. Their hosts include the fishes at the base of the food chain, the algal grazers and plankton feeders. Some of these, for example the parrotfishes, are important in the degradation of living and dead coral and in the formation of coral sand. Others eat various coral reef invertebrates and control their population densities. The pufferfishes (tetraodontids) and triggerfishes (balistids) may be im-

Slender suckerfish *Echeneis naucrates*. Note the sucking disc (a modified dorsal fin), protruding lower jaw, and distinctive striped coloration. Photo by D. Terver, Nancy Aquarium, France.

A marine leech from the buccal cavity of a tiger shark. The suckerfish and pilotfish may remove and eat this type of parasite. (Knuckle Reef)

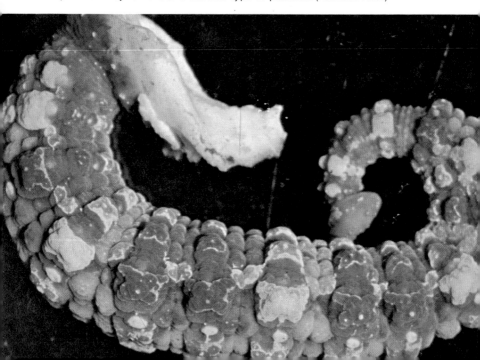

A blue streak *Labroides dimidiatus* soliciting one of the smaller fishes on the reef, a yellow damselfish *Pomacentrus moluccensis*. (2m., Lizard Island)

The juvenile *Labroides dimidiatus* lack the white bands of the adult. (5m., Lizard Island)

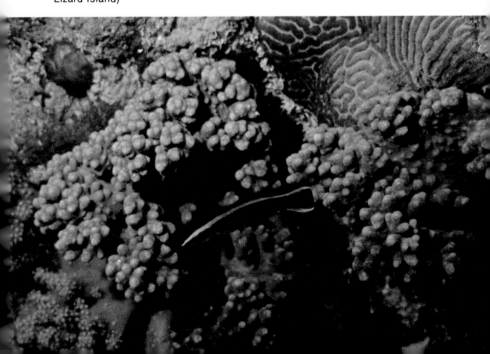

portant predators of the crown-of-thorns starfish, *Acanthaster planci*, which has devasted certain Indo-Pacific coral reefs. Other host fishes include the large reef predators, the snappers, groupers, scorpionfishes, trevallies and barracudas which feed on the smaller fishes.

Man himself is both indirectly and directly affected. Fishermen find that cleaning stations where fish congregate to be cleaned are productive fishing grounds. Sunken wrecks and artificial reefs are colonized by reef fishes, including the cleaners, and other fishes visit them to be cleaned.

Cleaning symbiosis therefore indirectly affects the entire coral reef community but four groups of organisms are directly involved: the cleaners (fishes and shrimps), the fishes cleaned (hosts or cleanees), the parasites removed and the mimics which imitate the cleaners.

CLEANERFISHES

By 1966, only a decade after Limbaugh's discovery of cleaning symbiosis, about 45 species of cleanerfishes had been identified.

Labroides dimidiatus, the most widespread cleanerfish of the Indo-Pacific. (3m., Lizard Island)

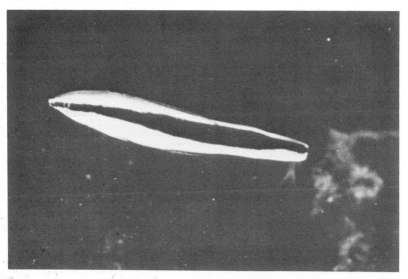

An adult *Labroides dimidiatus* grooming the surgeonfish *Acanthurus triostegus* in an aquarium. Photo by Miloslav Kocar.

Dozens more have been added to the list since then. Although the cleanerfishes belong to several different families most are remarkably similar in shape, size and behavior, an example of convergent evolution.

Morphology

All the cleaners, with the exception of the special cleaners of the large pelagic fishes, are small. Only a small fish can enter the mouths and gills of the majority of fishes or gain access to the other nooks and crannies where the ectoparasites often live. Only the young of a larger species of fish or the adults of a small species meet this requirement.

The second prerequisite is that the fishes must have a terminal mouth and suitable teeth to nip off the parasites. Fishes with suitable mouths include the gobies, butterflyfishes, wrasses and

Activity at a cleaning station over a 10-minute period. A soldierfish leaves as a parrotfish poses head down and fins splayed as a blue tang *Paracanthurus hepatus* waits. As the parrotfish departs the cleaners inspect the posing blue tang, pecking at parasites, mucus and/or scales.

Pufferfish *Arothron hispidus* enjoy the attention of *Labroides pectoralis.* The pleasure involved ensures that the host will continue to visit the cleanerfish even when it is in good health. Photo by Allan Power. (Great Barrier Reef)

damsels, and species of these fishes have become cleaners. A third prerequisite is that the fishes must eat small crustaceans, the most common ectoparasites of fish. The cleaners should also be conspicuous so that they are readily identifiable by their customers.

Guild signs

The cleaners characteristically have a colorful uniform which acts as a badge to identify them. This is called a guild sign and is analogous to the red and white striped pole which identifies a barber (a human cleaner) in so many countries, the traditional striped apron of the butcher or the black of the cleric or undertaker.

Cleaning behavior crosses the ocean's boundaries. This Caribbean cleaner, the neon goby *Gobiosoma oceanops,* grooms the Indo-Pacific batfish *Platax orbicularis.* Both fishes behave as if it were the most natural thing to do. Photo by H. Hansen.

The most common and widespread guild sign of the cleaner-fishes is longitudinal dark stripes on a brighter, often electric blue, background. Some of the specialized Indo-Pacific cleaner wrasses (*Labroides*), the neon gobies (*Gobiosoma*) of the Caribbean, certain Caribbean wrasses (*Thalassoma*) and juvenile combfish (*Coris*) of the Indo-Pacific have this coloration.

The similarity of the guild sign cannot be a coincidence, especially as fishes from the Atlantic recognize the status of Pacific cleaners in aquaria and vice versa. It is thought that the guild sign is very ancient, having evolved before the great oceans separated.

Cleaning stations

The cleanerfishes are conspicuously colored, but because they are small they must make their presence better known. All do this by living at prominent places, known as cleaning stations, which fish can readily identify. A station might be a large head of coral, a particularly prominent thicket of staghorn coral or an obvious cave. They are often on a route taken by fishes as they move over the reef according to the tide or time of the day.

An experienced diver can readily anticipate where a cleaning station may be. It may be anything conspicuous, large or unusually colorful or beautiful *which attracts his attention.*

Soliciting

The cleanerfishes may actively seek customers by approaching them or even chasing them. This soliciting may be performed with such persistence that a reluctant fish will attack the cleaner to drive it away. (I call to mind similar behavior by streetcorner hustlers.)

Some of the *Labroides* may perform an enticing little dance to attract customers. They swim, or perhaps more accurately 'flutter' back and forth in a jerky up-and-down manner

The soliciting may be performed by the host fish or cleanee. Fish already on a cleaner's clientele list will voluntarily present themselves at a cleaning station and may even jostle each other, vying for the cleaner's attention. Often only a single fish of a school will wish to be cleaned and will signal this by a ritualized

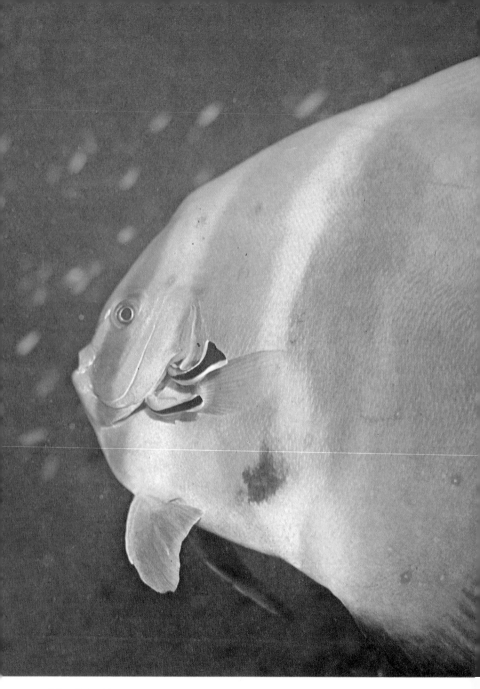

A batfish, *Platax orbicularis,* opens its operculum so that these blue streak cleaner wrasses can clean its gils. Photo by Walt Deas. (Heron Island)

Plagiotremus rhinorhynchos is not an exact mimic of the blue streak *Labroides dimidiatus*. It lures naive fishes to within striking distance and then launches a sudden attack, tearing off fatty eye tissues from its victim. Photo by Dr. Gerald R. Allen. (7m., Palau)

An almost exact copy of *Labroides dimidiatus* is this false cleaner *Aspidontus taeniatus*. It behaves like a cleaner to lure unsuspecting fishes within striking range. Photo by Dr. Herbert R. Axelrod. (Marau, Br. Solomon Islands)

color change and posing. Presumably it is that fish which leads the entire school to the cleaning station. Critically sick and wounded fishes have been seen presenting themselves to the cleaners time and time again until they succumbed to their injuries.

Hostfish

The cleaner's customers belong to most families of fishes resident on coral reefs or in surrounding waters. For example, hosts of the blue-streak wrasse include the small abundant damselfishes (Pomacentridae), the parrotfishes (Scaridae), other wrasses (Labridae), the large predatory groupers (Serranidae), butterflyfishes (Chaetodontidae), angelfishes (Pomacanthidae), snappers and emperors (Lutjanidae, Lethrinidae and Gaterinidae), trevallies (Carangidae) and many others. The blue-streaks have also been seen cleaning turtles.

Labroides dimidiatus may set up a cleaning station. There several fishes may be waiting for the resident cleaner(s) to finish those ahead of them in line.

Unlike the cleaners, the hosts have no obvious morphological adaptations to being cleaned but their behavioral adaptations are elaborate and complex. These adaptations, arising from natural selection over millennia, indicate that cleaning symbiosis is critical to the survival of the hosts. Fishes which did evolve the behavioral signals to attract cleaners had a survival advantage over those which did not. One can imagine that in severe epidemics of ectoparasites the fishes cleaned by the cleaners would survive and those not cleaned might die. Even if they did not die they would certainly produce fewer eggs than the healthy ones and their genes would therefore be selected against.

HOST-CLEANER BEHAVIOR

The fish wishing to be cleaned because it is diseased, wounded or simply enjoys the sensation of being groomed seeks out a cleaner at a cleaning station. Alternatively, it may be waylaid by a soliciting cleaner.

The host communicates its desire to be cleaned by changing color, opening its gill covers and mouth, spreading its fins to make ectoparasites accessible and assuming an inert motionless pose as if dead. The characteristic repertoire is recognized by the cleaner-fishes although it varies from species to species. Some fishes might change color completely, others might flush to a lighter or darker shade, some show spots or stripes. The pose depends on the centers of gravity and buoyancy of the inert fish. Some float head down, others tail down, others tilt to one side.

The cleanerfish then closely inspects the posing fish, which seems to enter a kind of trance. It pays special attention to the mouth, teeth, gills and fins, removing ectoparasites and food scraps and cleaning injured tissue. A blue-streak has been seen trying to pick off the warty protuberances which adorn the skin of a toadfish, and it is suspected that this and similar ornamentation might serve to attract the special attentions of the cleaners.

Tactile Reward System

The host finds the picking, nudging and caressings of the cleaner very pleasing. I suspect that the giant manta described

This parasitic isopod hides in the nares of the coral cod *Plectropomus maculatum,* possibly to avoid the attentions of the cleanerfishes. (Lodestone Reef)

Blue streak cleanerfish *Labroides dimidiatus* and a sinuously swimming mimic *Plagiotremus rhinorhynchos* (upper fish). The cleaner had just finished cleaning the mimic. (10m., Lizard Island)

The juvenile *Coris picta* also engages in cleaning behavior. Note its strong resemblance to the juvenile blue streak cleaners. Photo by Rudie Kuiter. (Sydney)

Young *Coris picta* wear the guild sign of the cleanerfishes and are actually cleaners. Here a *Coris picta* is cleaning two mado *Atypichthys strigatus*. Photo by Wade Doak (New Zealand)

earlier was addicted to it as it did not seem badly infected by ectoparasites.

The enjoyment factor is described by ethologists as a "tactile reward system" and it ensures that the hosts will present themselves for cleaning even when they are free of ectoparasites. If they presented themselves only when they were sick the cleaner-fishes would starve in times of low incidence of parasites. At these times the cleaner removes scales, mucus and food scraps from healthy fish—it becomes a commensal or even a parasite, but little harm is done.

Immunity From Predation

The full-time cleanerfish can approach and even enter the mouths of predatory fishes such as barracudas, groupers and snappers although the diets of these fishes often include prey of a similar size to the cleanerfish. This does not occur because the cleanerfish is toxic or poisonous, as the flesh of a dead cleanerfish is readily eaten. The selective advantages of this suppression of the feeding drive is obvious. If the predators ate the fish which cleaned them they would never have their parasites removed.

Cleanerfishes are not entirely immune from predation by the reef fishes. One was observed being eaten by the bottom-dwelling lizardfish *Synodus* (seen here). Photo by Dr. Herbert R. Axelrod. (Marau, Br. Solomon Islands)

Again the great importance of the phenomenon of cleaning is indicated.

The part-time cleaners, fishes which clean only to supplement their diets, enjoy only a partial immunity—they are sometimes eaten by the host fish. Although no blue-streak cleanerfish have ever been found in stomach contents of coral reef fishes they are not completely immune from predation as a colleague once saw one being taken by a bottom-dwelling lizardfish (*Synodus*). This fish and some of the sedentary scorpionfishes are poor swimmers and probably would not be capable of swimming considerable distances to be cleaned. Indeed, the highly venomous stonefish (*Synanceja*) move little and may actually encourage the settlement of algae and other organisms to perfect their camouflage. The common belief in the immunity of full-time cleaners to predation should be qualified: they are immune only to *those species of fishes which they clean.*

Termination Of Cleaning

When the cleanerfish has finished cleaning it swims away from the host and returns to the cleaning station. The host awakens from its trance, assumes its normal posture and color and swims off.

Alternatively, if the cleaning process is lengthy or painful or if a posing host is disturbed by the approach of a predator, then it will become increasingly uneasy or 'fidget y.' The cleaner may nudge or jab it to pacify it (again this behavior has human connotations). If this is unsuccessful and the host desires to terminate the cleaning it will signal this to the cleaner. A small jerk or a snap of its jaws is the usual signal for the cleaner to leave immediately. Sometimes a startled fish will not have time to do this. A biologist once saw a fleeing trevally or jack hiccup and spit out a rather confused little cleaner.

SPECIES OF CLEANERFISHES

It is not possible to describe all the known cleanerfish species, but some of the better known and better studied species merit a specific discussion.

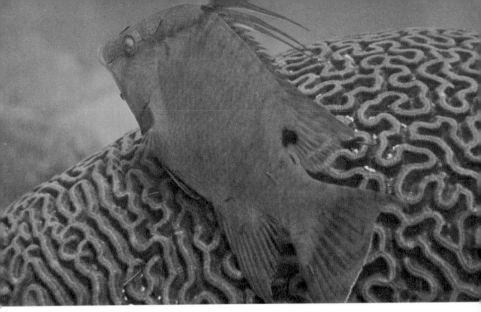

A host of neon gobies *Gobiosoma oceanops* simultaneously cleaning different parts of the hogfish *Lachnolaimus maximus*. The hogfish will change color making variously colored parasites more visible. Photo by Dr. Walter A. Starck II. (Florida)

An unknown clingfish *Lepadicththys* sp. browsing on a moray eel *Gymnothorax prasinus*. Photo by Walt Deas. (Sydney Harbor)

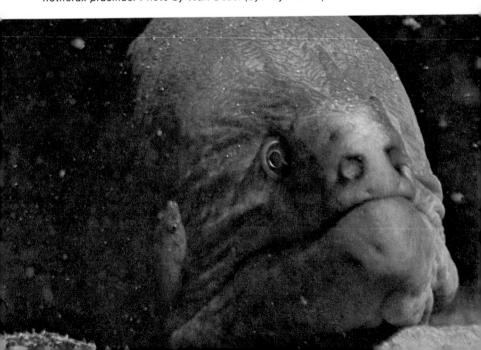

Cleaning gobies may clean large predatory fishes without being eaten. Here *Gobiosoma evelynae* grooms the tiger grouper *Mycteroperca tigris*. Photo by Carl Roessler. (Netherlands Antilles)

An adult gray angelfish *Pomacanthus arcuatus* poses for the cleaner goby *Gobiosoma illecebrosum.* Photo by Carl Roessler. (Islas Rosarios, Colombia)

Caribbean Cleanerfishes

The vividly colored neon gobies are the best known cleaners in the Caribbean. There are twelve species associated with corals or sponges, and six of these are fish cleaners. The best known is *Gobiosoma (Elacatinus) oceanops,* which bears the cleaner guild sign and lives on coral heads which act as the cleaning station. Host fishes come to the coral and pose. The goby, a weak swimmer like most of its family, moves by skipping from surface to surface, holding on by means of a sucker formed from the pelvic fins, and jumps onto the host's body. It skips around on it inspecting its scales, mouth, gills and fins for parasites and wounds. Its sucking and skipping seem to add to the pleasure of the cleaning. A juvenile of the wrasse *Thalassoma bifasciatum* may also live at the cleaning station of *G. oceanops* and other *Gobiosoma* species of cleanerfishes.

Several neon gobies, *Gobiosoma oceanops,* wait on a coral head for the next fish to come and be cleaned. Photo by Dr. Walter A. Starck II. (Florida)

The Indo-Pacific cleaner wrasse cleaning the Caribbean Spanish hogfish *Bodianus rufus*. Note the similarity in pattern to the cleaning goby *Gobiosoma oceanops* opposite. Photo by K. Paysan.

George Losey, an authority on cleanerfish behavior, studied the association between this wrasse and gobies while living in an underwater habitat in 20 m. of water off Puerto Rico. He suggests that the relationship between the two cleaners is mutualistic: the wrasse, being much larger than the goby, is more conspicuous and attracts more fish while the goby, because of its sucking, provides added pleasure ensuring their return.

The Spanish hogfish, *Bodianus rufus*, is another well known Caribbean cleaner, living in the deeper water at the reef edge where it is frequently visited by pelagics as well as local reef fishes. It is not completely immune from predation and is a favorite on the menu of the trumpetfish, which approaches its prey by hiding behind a fish waiting to be cleaned. When the hogfish nears the host to clean it the trumpetfish darts out from behind and eats it.

Indo-Pacific Cleanerfishes

The specialized cleanerfishes of coral reefs of this region are the wrasses, *Labroides* spp. They are full-time cleaners and rarely, if

A pair of cleaning shrimp *Lysmata grabhami* grooming the head of a moray eel *Gymnothorax* sp. A juvenile cleanerfish *Labroides dimidiatus* hovers nearby. Photo by Dr. John E. Randall.

The guild sign of the cleaner shrimps seem to be the long white antennae, clearly seen in this *Lysmata grabhami* and *Stenopus hispidus* (opposite page). Photo by Rodney Jonklaas. (Sri Lanka)

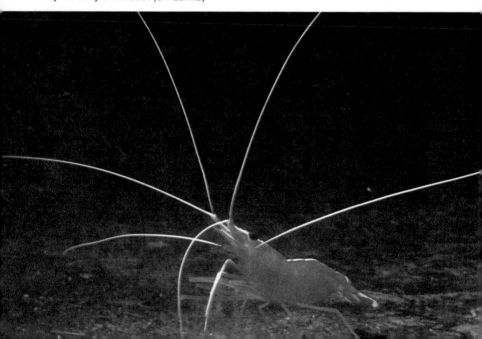

A pair of banded coral shrimp *Stenopus hispidus* wave their conspicuous white antennae from their shelter to attract passing fishes. These shrimp are part-time cleaners and are eaten by a number of different reef fishes. (5m., Kangaroo Reef)

The long chelae of the banded coral shrimp are useful in removing the parasites. Photo by Dr. Patrick L. Colin. (15m., Discovery Bay, Jamaica)

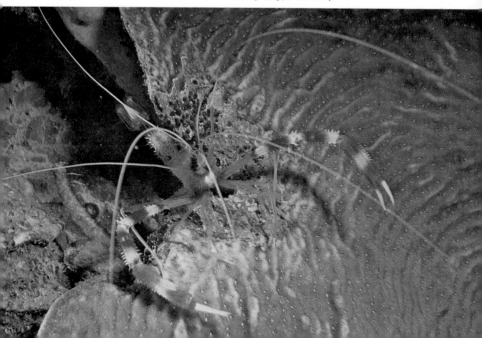

The resident cleaner wrasse of the Hawaiian Islands *Labroides phthirophagus.* Photo by Dr. Herbert R. Axelrod.

ever, take any other food. There are many other cleaners in this region but they are poorly known, undoubtedly eclipsed by the *Labroides.*

The blue-streak (*L. dimidiatus*) is the best known and most widespread species of *Labroides.* It has the distinctive guild coloration, a soliciting dance, maintains a cleaning station and is strongly territorial. Like other wrasses and parrotfishes it changes its sex during its life. Its social order is unusual in that a single male dominates a harem of females, within which there is a hierarchy or peck order. If the male is removed or dies the dominant female changes into a male and takes over. If an adjacent male steals the harem during the change-over process, the new male reverts to a female.

The gray and yellow *L. bicolor* often lives on the same reefs as the blue-streak but favors more exposed areas. It is a larger fish and therefore does not clean small fishes or enter the mouths of larger fishes. It does not do the soliciting dance and its territory is much larger than that of the blue-streak.

The Hawaiian *L. phthirophagus* is more like the blue-streak but is confined to those islands, is slightly smaller and rarely dances. It cleans its host under cover, for example under a ledge or in a cave.

The small *L. rubrolabiatus* also has a limited distribution, the southeastern Pacific coral reefs. It overlaps with the blue-streak and *L. bicolor* in the Society Islands, where all three have been seen cleaning the same host and cleaning each other.

No doubt there are many other cleaner fishes in this region but few are well known. Juveniles of the sub-tropical combfish wrasse (*Coris picta*) closely resemble the blue-streak and clean other fish.

While researching this chapter I began to look for other examples of cleaning symbiosis on the fringing reefs near the Great Barrier Reef and found that juveniles of two species of *Halichoeres* on the reefs near my house on Magnetic Island are also cleaners. One lives near a blue-streak cleaning station and cleans fishes which settle on the bottom while being cleaned by the blue-streak, and the other chases after a host as it swims near. Neither carry the guild sign and fish do not pose for them. Both spend a large proportion of their time foraging for food in algal beds. I also discovered that a third species, probably a juvenile butterfish or scat, cleaned the fish in a nearby estuary. Hosts such as archer-

Labroides bicolor may occur along with the blue streak *L. dimidiatus* but behaves somewhat differently from that species. Photo by Dr. Herbert R. Axelrod. (Maldive Islands)

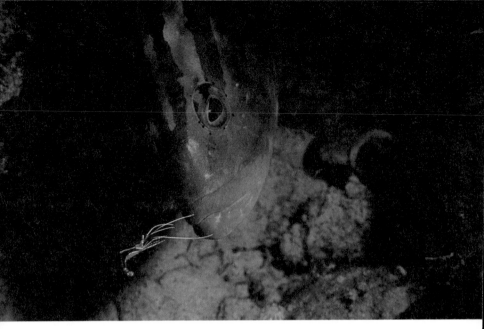

Pederson's cleaner shrimp *Periclimenes pedersoni* grooming the lips of the Nassau grouper *Epinephalus striatus.* Photo by Dr. Patrick L. Colin. (Whale Cay, Bahamas)

Periclimenes pedersoni on their host anemone, the cleaning station. Note the characteristic white antennae, the guild sign of the cleaner shrimps. Photo by Dr. Dwayne Reed. (Bonaire)

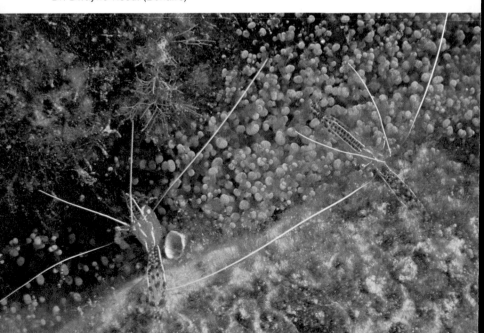

fishes (*Toxotes*), mullets (*Mugil*) and long-toms (*Tylosurus*) visited the station, a boat mooring, and posed for cleaning before the centimeter-long fish.

CLEANER SHRIMPS

Conrad Limbaugh also discovered that certain species of shrimps clean fishes. The symbiosis resembles that of the cleaner-fishes: the shrimps are conspicuously marked, solicit passing fishes, live at cleaning stations, the host fishes pose, parasites and infected tissue are removed using the shrimps' long nippers or chelae and they enjoy a certain degree of immunity from predation. Most of the known cleaner shrimps are Caribbean species, but undoubtedly many await discovery in the Indo-Pacific.

Least specialized of the cleaner shrimps is the Californian cleaner, *Lysmata californica*. It is not as conspicuously marked as the others and instead of soliciting customers it cleans only those fishes which happen to shelter in a crevice where it lives. It cleans to supplement its diet and as only a part-time cleaner it is on the diet of many fishes. Certain fishes placed in aquaria with these shrimps promptly eat them, but there are cases of the shrimps eating the fishes. Obviously the symbiosis is not refined.

The related Grabham's shrimp (*L. grabhami*) is more specialized. It is brightly colored with long white tentacles which it waves at passing fish. Unlike the California cleaner, it relies on cleaning for most of its diet and is therefore immune from predation by many fishes.

Grabham's shrimp often lives in association with another cleaner shrimp, the banded coral shrimp, *Stenopus hispidus*, a most beautiful reef dweller. This shrimp lives in sexual pairs, a smaller male living near or often on the larger female. It has long white antennae which protrude from the security of its shelter. It is a part-time cleaner and on the diet of many predators. Both the banded coral shrimp and Grabham's shrimp have a tropical Atlantic and Indo-Pacific distribution, but another smaller *Stenopus* lives only in the Caribbean. All of these shrimps clean their hosts without leaving their shelter.

Pederson's shrimp, *Periclimenes pedersoni*, is an elegant little cleaner which lives among the tentacles of large Caribbean sea

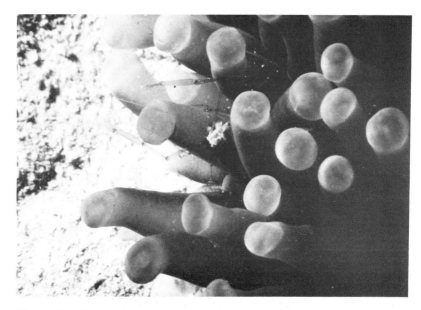

The coral shrimp *Periclimenes holthuisi* is a part-time cleaner of fishes which visit the mushroom coral *Fungia actiniformis*. (10m., Lizard Island)

anemones. It also has long white antennae which it seductively lashes at passing fishes. A fish desiring to be cleaned will pose near the anemone, the shrimp's cleaning station and refuge. The shrimp leaves the anemone and jumps onto the skin of the host, inspecting it for parasites which it removes with its sharp chelae. Pederson's shrimp may even make incisions in the host skin to extract subcutaneous parasites. When it is finished attending to the fish the shrimp crawls off it and returns to the anemone. The neon goby often maintains a cleaning station near the anemone and the two clean the same hosts.

The Yucatan shrimp, *P. yucatanicus*, also lives in Caribbean anemones and has long white antennae which are lashed back and forth to attract fishes. However, Limbaugh suspected that this shrimp might not be a cleaner—it might be a cleaner-shrimp-mimic feeding on host tissue.

The larger anemones of the tropical Indo-Pacific harbor similar shrimps, for example *P. holthuisi* and the very beautiful *P. brevicarpalis*, but until recently there was no evidence that any of them were cleaners. However, colleagues and I have on several occasions seen groups of *P. holthuisi* crawling over the coral cod *Plectropomus*, presumably cleaning them. The fish lie on the shrimp's home, the mushroom coral *Fungia actiniformis*, and the shrimps then climb onto them.

A bottom-dwelling Indo-Pacific shrimp, *Leandrites cyrtorhynchus*, is now known to be a cleaner often living near blue-streak cleaning stations. It is an inconspicuous shrimp and relies on those fishes attracted by the blue-streak. Hosts give it a similar signal, such as mouth closing, to indicate that they wish to leave. This shrimp may clean the bottom-dwelling fishes which do not swim large distances to a cleaning station.

The very attractively banded *Stenopus hispidus*. The chelae or pincers used to remove their host's parasites are clearly visible. Photo by R.P.L. Straughan. (Aquarium photo)

ECTOPARASITES

The third group of animals directly involved in cleaning symbiosis are the ectoparasites which are removed and eaten by the cleaners.

Crustaceans are the most common fish ectoparasites. Most are very small, a millimeter or two in length, but a few are much larger. Copepods (Calanidae, Lernaeidae and Caligidae) and isopods (Gnathiidae and Cymothoidae) are the most common crustaceans living on the body surface and in the gills. A large number of trematode parasites also live in the gills of reef fishes. Marine leeches parasitize sharks' body surfaces, mouths and gills. Fungi frequently cause diseases of the skin or occur as secondary infections in wounds. All receive the attention of the cleaners.

It has been suggested that cleaning has affected the evolution of crustacean parasites as many are cryptically colored. Some are pigmented to match their host's color and others are transparent, possibly to escape the attentions of a cleaner. Perhaps the color change by the host fish during cleaning acts to make the parasite conspicuous as well as acting as a signal to the cleaner. In the North Sea where there are reputed to be few cleaners the ectoparasites are more obvious than those of tropical seas.

CLEANER MIMICS

There is a villain in all melodramas, and the story of the cleaners is no exception. The culprits are small fishes, the sabre-toothed blennies, which mimic the guild sign and behavior of the blue-streak, the model.

The sabre-toothed blenny *Aspidontus taeniatus* is an almost perfect copy of the blue-streak except that it does not remove the parasites of the host. *Aspidontus* lives near cleaning stations where, like the true cleaner, it dances up and down trying to attract fish. A naive fish will present itself to *Aspidontus*, spreading out its fins and opercula, no doubt anticipating the pleasures of cleaning. Imagine the shock it experiences when the 'cleaner' attacks it, tearing off a piece of fin.

Just as the pleasure of the tactile reward system acts to reinforce the host-cleaner bond, the trauma of the attack acts to reduce it.

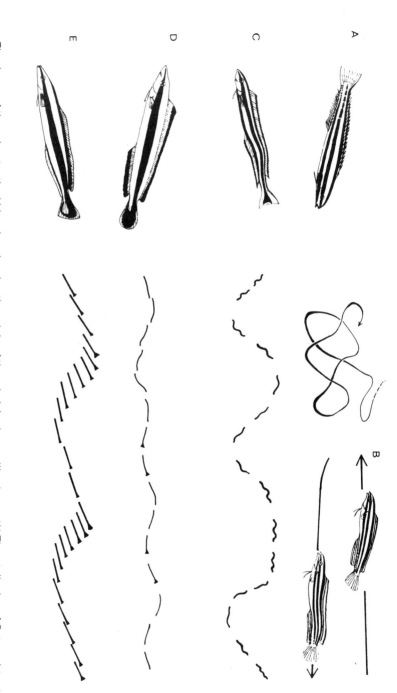

The dance of the sabre-toothed blennies showing the origin of the mimic's cleaner-like dance. (A) Threat display of *Petroscirtes temminckii*. (B) Courtship dance of *P. temminckii*. (C) Threat display and dance of *Plagiotremus rhinorhynchos*. (D) Threat display and dance of the mimic *Aspidontus taeniatus*. (E) Dance of the cleaner wrasse *Labroides dimidiatus*. (Redrawn from Wickler)

Fishes repeatedly attacked may become wary of all cleaners, but generally an adult fish learns to detect the mimic. Territorial fishes learn to avoid the area of the attack and, as the mimic is strongly territorial, they therefore avoid it even though they never actually learn to distinguish model from mimic. Fishes may also detect discrepancies in the mimic's color, shape and behavior, and posing fishes become suspicious of any 'cleaner' which approaches from the rear—the direction the mimic attacks from.

Consequently there are strong evolutionary pressures for the mimic to more closely resemble the model, as the more accurate replicates would delude more fish than the less accurate ones. Wolfgang Wickler, the behaviorist who described this mimicry, considers that *Aspidontus* is one of the most exact mimics known. Even racial variations of the true cleaner are copied: in the Tuamotu Archipelago the blue-streaks have an unusual orange blotch which the local mimics have copied.

Strong evolutionary pressures have caused the mimic to more closely resemble the model. *Aspidontus taeniatus* has been considered one of the most exact mimics known. Photo by Kok-Hang Choo. (Taiwan)

The grinning bandit, the cleaner fish mimic *Plagiotremus*, usually can be seen peering from its lair, an empty snail tube on a giant clam or coral head. When out like this it is usually on the attack. (3m., Lizard Island)

A closely related sabre-toothed blenny, the blue-lined blenny (*Plagiotremus rhinorhynchos*), also mimics the blue-streak, but it is at best a poor imitation and deceives only young or very naive fishes. This blenny is more slender than the blue streak but with an erected dorsal fin its proportions are vaguely similar. It is a jerky swimmer, but this action looks remotely like the blue-streak's dance.

The blue-lined blenny lives in a hole near a cleaning station and peers out, looking rather like a cunning masked bandit plotting an ambush. When it is planning to attack it changes its color from black and blue-gray stripes to black and bright blue. Its ruse may be far from perfect, but it does enable the blue-lined blenny to get close enough to a fish to launch a lightning attack, tearing off pieces of the victim's fatty eye tissue. The attack must be very painful—I have seen relatively large trevally and an adult parrot-fish frantically fleeing the scene of an ambush.

Plagiotremus represents an intermediate stage in the evolution of the mimicry. We can trace its evolution back even farther to the less specialized sabre-toothed blennies such as *Petroscirtes* and *Dasson* which resemble ancestors which were predisposed to the evolution of the mimicry.

Petroscirtes and *Dasson* are small, coincidentally a size similar to the cleaner, and they also have longitudinal stripes. They are poor swimmers, for fishes of the family Blennidae have lost their swim bladders to become bottom-dwellers. The sabre-toothed blennies

The cleanerfish mimic changes color according to its state of motivation. (A) Frightened and driven to defense. (B) Confident and prepared to attack: the coloration which resembles that of *Labroides dimidiatus*. (C) Actively fighting. (D) Greatly frightened. (Redrawn from Wickler)

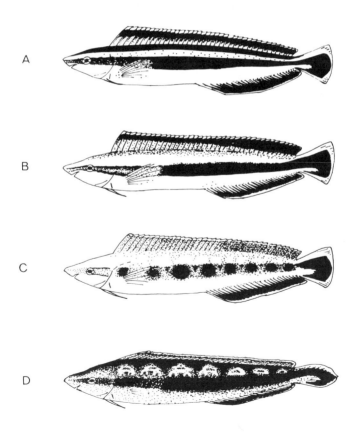

have reacquired swim bladders but they are still poor swimmers and rely heavily on their pectoral fins for propulsion. The wrasses also rely heavily on a pectoral beat, so the mode of swimming of wrasse and sabre-tooth blenny is superficially similar. One other factor has preadapted the sabre-toothed blennies to the mimicry: they can change their colors during intra-specific displays and one color approximates the guild sign. Natural selection has perfected this accident.

It is thought that the dance of the cleaner mimic evolved from the ritualized head bobbing, fin displaying and color changing of the ancestral blenny. However, this territorial and courtship behavior has lost all its meaning in *Aspidontus* for it bobs its head to any object it sees, even as it swims along. The sum total of this jerky swimming and head bobbing is a dance remarkably like that of the blue-streak cleaner.

CONCLUSION

The amazing story of the mimics seems an appropriate conclusion to the saga of the cleaners and the symbionts in general. To me cleaning symbiosis, its critical importance in the ecosystem and the exploitation by the mimics epitomizes the beauty and complexity of Nature. In it we see evolution in action, plots and counterplots, and we can identify with the participants. The episode with the manta one gray day on a coral reef gave me an insight into the importance of the symbiosis and the intricacies of Nature. Another experience reinforced it.

This time it was in a sub-tropical sea about 1000 km. south of the Great Barrier Reef in a unique zone of tropical, temperate and cold water biota. This day was one in a thousand. A deep blue sea mirrored a cloudless summer sky. Underwater visibility was superb as I swam through rocky grottos decorated by trees of black corals and gorgonians, and the coral reef vibrant with life. Even the most timid species shamelessly displayed for me. Turtles, lionfishes, Moorish idols, angelfishes and parrotfishes from the tropics mingled with mado, bream, morwong and butterfish from the temperate south. Evidently tropical parasites also mingled with their temperate counterparts as cleaners were at work everywhere I looked.

In an area not much larger than a house I saw almost every known cleaner of the tropical and temperate South Pacific. Bluestreaks danced up and down outside their stations; juveniles of two species of combfish, temperate cleaners, groomed tropical parrotfishes. Banded coral shrimps seductively waved their long white antennae from the security of crevices. Blue-lined blennies relentlessly searched for gullible prey and cleaner mimics expertly deceived visitors, including myself.

On that day I had a strange, almost mystical experience, perhaps induced by nitrogen narcosis following several deep dives. I left my slate and camera in the boat and instead of making notes of what I saw, I experienced it. I felt a part of it, just another of the little fishes—I was no longer an intruder. I found myself humbly musing: *How inflated must man's ego be if he thinks he can ever understand this? Why must he always seek to explain Nature?* But my air ran out and I had to return to the surface.

APPENDIX:
PHOTOGRAPHIC TECHNIQUES

TECHNIQUES

Special emphasis has been placed on the photographic material as the subject material is often very colorful and would be unfamiliar to many readers. The illustrations attempt to convey the morphological specializations of the symbionts, their coloration and camouflage, and something of their behavior and habitats.

Where possible the photographs are of living subjects in their natural surroundings. Subjects were firstly photographed and then collected for aquarium close-ups and preserved for identification. For large subjects underwater the author used an Asahi Pentax and 20mm. lens plus correcting dome port. For small subjects a 55mm. macro-lens was used. Magnification of X2 could be obtained using a set of extension tubes with this lens. An electronic flash was used for illumination. Equipment was housed in the versatile Ikelite system. A Nikonos amphibious camera was used as a backup camera and a Zenza Bronica 6x6 camera in a P.M.M. housing was used for special subjects.

Walter Deas used a Rollei SL66 and a Rolleimarin 6x6 camera for the cleanerfish photographs. The other photographers used similar 35mm. and 6x6cm. equipment.

Very small subjects were photographed through a Carl Zeiss photomicroscope, and larger subjects, between 0.5mm. and 10mm., were photographed with a Carl Zeiss Tessovar lens with x0.8 - x25.0 zoom optics. Prolonged exposures of between 5 secs. and 2 mins. were needed for maximum depth of field. Subjects were usually chilled for a few minutes prior to photographing to reduce movement. Kodak Ektachrome-X and Kodachrome 64 were used throughout, above and below water. Most of the illustrations were taken on the Great Barrier Reef of Australia. Others were taken in southern, western and northwestern Australia, and in New Guinea to the north.

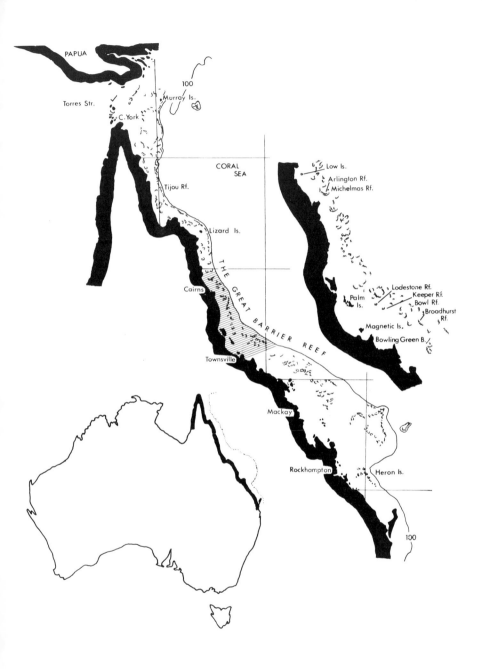

The Great Barrier Reef of Australia showing the main photographic localities.

BIBLIOGRAPHY
(SELECTED REFERENCES, CHAPTER BY CHAPTER)

Introduction and Chapter One

Caullery, M. (1952). *Parasitism and Symbiosis*. Sidgwick and Jackson, London.

Cheng, T.C. (1967). Marine molluscs as hosts for symbioses. *Adv. mar. Biol.*, **5:** 1-424.

DeBary, A. (1879). Die Erscheinung der Symbiose. Cassel, Strasburg.

Gotto, R.V. (1969). *Marine Animals. Partnerships and Other Associations*. The English Universities Press Ltd., London.

Henry, S.M. (1966). *Symbiosis*. Vol. 1. Academic Press, New York.

MacGinitie, G.E. and MacGinitie, N. (1949). *Natural History of Marine Animals*. McGraw-Hill, London.

Chapter Two

Baer, J.C. (1971). *Animal Parasites*. World University Library, London.

Cheng, T.C. (1967). Marine molluscs as hosts for symbioses. *Adv. mar. Biol.* **5:**1-424.

Dogiel, Petrushevski and Polyanski (1970). *Parasitology of Fishes*. T.F.H., New Jersey.

Meyerhof, E., and Rothschild, M.A. (1940). A prolific trematode. *Nature, London*, **146.**

Sindermann, C.J. (1970). *Principal diseases of marine fish and shellfish*. Academic Press, New York.

Chapter Three

Castro, P. (1971). Nutritional aspects of the symbiosis between *Echinoecus pentagonus* and its host in Hawaii, *Echinothrix calamaris*. **In**: *Aspects of the Biology of Symbiosis* (Cheng (ed.)). University Park Press, Baltimore.

Crisp, D.J. and Meadows, P.S. (1962). The chemical basis of gregariousness in Cirripedes. *Proc. Roy. Soc.* **156B**:500-520.

Crisp, D.J. (1965). Surface chemistry, a factor in the settlement of marine invertebrate larvae. *Botanica Gothoburgensia* **iii**:51-65.

Davenport, D. (1955). Specificity and behaviour in symbioses. *Quart. Rev. biol.* **30**:29-46.

Dimock, R.V. (1974). Intraspecific aggression and the distribution of a symbiotic polychaete on its host. **In**: *Symbiosis in the Sea*. (Vernberg (ed.)). University of South Carolina Press, Columbia.

Portier, C. (1918). *Les symbiotes*. Paris.

Ross, A. & Newman, W.A. (1969). A coral-eating barnacle. *Pac. Sci.* **23**:252-256.

Zann, L.P. (1975). Biology of a barnacle (*Platylepas ophiophilas* Lanchester) symbiotic with sea snakes. **In:** *The Biology of Sea Snakes*. (Dunson (ed.)). University Park Press, Baltimore.

Chapter Four

Barry, C.K. (1965). Ecological study of the decapod crustaceans commensal with branching coral *Pocillopora meandrina* var. *nobilis* Verrill. M.S. Thesis. University of Hawaii, Honolulu.

Benson, A.A. (1975). Energy exchange in coral reef ecosystems. *Reef Biogenesis Symposium*, Dec. 1975. Australian Institute of Marine Sciences.

Bosch, H.F. (1965). A gastropod parasite of solitary corals in Hawaii. *Pac. Sci.* **19**: 267-268.

Bruce, A.J. (1975). Coral reef shrimps and their colour patterns. *Endeavour* **34**: 23-27.

Cloud, P.E. (1959). Geology of Saipan, Mariana Islands. 4. Submarine topography and shoalwater ecology. *U.S. Geol. Surv. Prof. Paper* **280-k**: 361-445.

Davis, W.P., and Cohen, D.M. (1968). A gobiid fish and a palaemonid shrimp living on an antipatharian sea whip in the tropical Pacific. *Bull. mar. Sci.* **18**: 749-761.

Ebbs, N.K. (1966). The coral-inhabiting polychaetes of the Northern (southeastern) Florida reef tract. I. Aphroditidae, Polynoidae, Amphinomidae, Eunicidae, and Lysaretidae. *Bull. mar. Sci.* **16**: 485-555.

Gohar, H.A.F., and Soliman, G.N. (1963). On the biology of three coralliophilids boring in living coral. *Publ. Mar. Biol. Sta-Al-Ghardaga, Red Sea.* **12**:99-126.

Gohar, H.A.F., and Soliman, G.N. (1963). On three mytilid species boring in living coral. *Ibid*: 65-98.

Goreau, T.F., Goreau, N.I., Sbot-Ryen, T., and Yonge, C.M. (1969). On a new commensal mytilid (Mollusca: Bivalva) opening into the coelenteron of *Fungia scutaria* (Coelenterata). *J. Zool., Lond.* **158**: 171-195.

Harris, L.G. (1971). Nudibranch associations as symbiosis. **In:** *Aspects of the Biology of Symbiosis.* Cheng (ed.)). University Park Press, Baltimore.

Hein, F.J. and Risk, M.J. (1975). Bioerosion of coral heads: inner patch reefs, Florida Reef tract. *Bull. mar. Sci.* **25**: 133-137.

Hiro, F. (1938). Studies on the animals inhabiting coral reefs. 11. Cirripeds of the genera *Creusia* and *Pyrgoma. Palao Trop. Biol. Stat. Studies* **3**:391-417.

Jaccarini, V. and Bannister W.H. (1968). The pallial glands and rock boring in *Lithophaga lithophaga* (Lamellibranchia, Mytilidae). *J. Zool., Lond.* **154**: 397-401.

Knudsen, J.W. (1967). *Trapezia* and *Tetralia* (Decapoda, Brachyura, Xanthidae) as obligate ectoparasites of pocilloporid and acroporid corals. *Pac. Sci.* **21**:51-57.

Neumann, A.C. (1966). Observations on coastal erosion in Bermuda and measurements of the boring rate of the sponge, *Cliona lampa. Limnol. and Oceanogr.* **11**: 92-108.

Newman, W.D. (1960). On the paucity of intertidal barnacles in the tropical Western Pacific. *Veliger.* **2**: 89-94.

Patton, W.K. (1966). Decapod crustacea commensal with Queensland branching corals. *Crustaceana.* **10**:271-295.

Patton, W.K. (1972). Studies on the animal symbionts of the gorgonian coral, *Leptogorgia virgulata* (Lamarck). *Bull. mar. Sci.* **22**:419-431.

Pang, R.K. (1973). The ecology of some Jamaican excavating sponges. *Bull. mar. Sci.* **23**:227-243.

Potts, F.A. (1915). *Hapalocarcinus*, the gall-forming crab, with some notes on the related genus *Cryptochirus. Papers Dep. Marine Biol. Carnegie Inst. Wash.* **8**:33-69.

Randall, J.E. (1967). Food habits of reef fishes of the West Indies. *Stud. trop. Oceanogr. Miami.* **5**:665-847.

Rhode, K. (1975). Ecological significance of parasites on the Great Barrier Reef. *Reef Biogenesis Symposium*, Dec. 1975. Australian Institute of Marine Sciences.

Robertson, R. (1970). Review of the predators and parasites of stony corals, with special reference to symbiotic prosobranch gastropods. *Pac. Sci.* **24**: 43-54.

Ross, A. and Newman, W.A. (1973). Revision of the coral-inhabiting barnacles (Cirripedia: Balanidae). *San Diego Soc. Nat. Hist. Trans.* **17**:137-174.

Stock, J.H., and Humes, A.G. (1969). *Cholomyzon palpiferum* n.gen., n.sp., a siphonostome cyclopoid copepod parasitic in the coral *Dendrophyllia* from Madagascar. *Crustaceana.* **16**: 57-64.

Stoddart, D.R. (1969). Ecology and morphology of recent coral reefs. *Biol. Rev.* **44**:433-498.

Strasburg, D.W. (1966). Observations on the ecology of four apogonid fishes. *Pac. Sci.* **20**:338-341.

Wells, J.W. (1957). Coral reefs. **In**: Treatise on marine ecology and paleoecology. *I.* Ecology. *Memoirs, Geol. Soc. Amer.* **67**: 609-631.

Yonge, C.M. (1955). Adaptation to rock boring in *Botula* and *Lithophaga* (Lamellibranchiata, Mytilidae) with a discussion on the evolution of this habit. *Quart. J. Micr. Sci.* **96**:383-410.

Yonge, C.M. (1967). Observations on *Pedum spondyloideum* (Chemnitz) Gmelin, a scallop associated with reef-building coral. *Proc. malac. Soc. Lond.* **37**:311-323.

Chapter Five

Colin, P. (1976). *Neon Gobies.* T.F.H., New Jersey.

Dembowska, W.S. (1926). Study of the habits of the crab *Dromia vulgaris* M.E. *Biol. Bull.* **50**: 163-178.

Goreau, T.F., and Yonge, C.M. (1968). Coral community on muddy sand. *Nature* **217**: 421-423.

Healy, A. & Yaldwyn, J. (1970). *Australian Crustaceans in Colour.* Reed, Sydney.

Hurley, A.C. (1973). Fecundity of the acorn barnacle *Balanus pacificus* Pilsbry: a fugitive species. *Limnol. and Oceanogr.* **18**: 386-393.

Karpus, I., Szlep, R., and Tsurnamal, M. (1972). Associative behaviour of the fish *Cryptocentrus cryptocentus* (Gobiidae) and the pistol shrimp *Alpheus djiboutensis* (Alpheidae) in artificial burrows. *Mar. Biol.* **15**:95-104.

Laubenfels, M.W. de. (1936). A discussion of the sponge fauna of Dry Tortugas and the West Indies in general, with material for a revision of the families of Porifera. *Carnegie Inst. Wash. Pap. Tortugas Lab.* **30**:(iii) 225 pp.

Moehring, J.L. (1970). Communication systems of a goby-shrimp symbiosis. *Western Society of Naturalists, Annual Meeting 51st.* ABSTS.: 24-25.

Patton, W.K. (1972). Studies on the animal symbionts of the gorgonian coral, *Leptogorgia virgulata* (Lamark). *Bull. mar. Sci.* **22**:419-431.

Pearse, A.S. (1950). Notes on the inhabitants of certain sponges at Bimini. *Ecology.* **31**:149-151.

Rosewater, J. (1969). Gross anatomy and classification of the commensal gastropod *Caledoniella montrouzieri* Souverbie, 1869. *Veliger.* **11**:345-350.

Tyler, J.C. and Bohlke, J.E. (1972). Records of sponge-dwelling fishes, primarily of the Caribbean. *Bull. mar. Sci.* **22**:601-642.

Chapter Six

Biggs, J.T., and Morejohn, G.V. (1972). Barnacle orientation and waterflow characteristics in Californian grey whales. *J. Zool.* **167**:287-292.

Bruce, A.J. (1972). An association between a pontoniinid shrimp and a rhizostomatous scyphozoan. *Crustaceana.* **23**:300-302.

Crisp, D.J. (1973). Factors influencing the settlement of marine invertebrate larvae. In: *Perspectives in Chemoreception by Marine Organisms* (Grant (ed.)). Academic Press, New York.

Dahl, E. (1963). The association between young whiting, *Gadus merlangus*, and the jellyfish *Cyanea capillata. Sarsia.* **3**:47-55.

Darwin, C. (1854). *A Monograph on the Sub-Class Cirripedia.* Cramer, Weinheim.

Gooding, R.M. and Magnuson, J.J. (1967). Ecological significance of a drifting object to pelagic fishes. *Pac. Sci.* **21**: 486-497.

Harris, L.G. (1971). Nudibranch associations as symbiosis In: *Aspects of the Biology of Symbiosis.* (Cheng (ed.)). University Park Press, Baltimore.

Heyerdahl, T. (1950). *The Kon-Tiki Expedition.* George Allen and Unwin Ltd., London.

Lane, C.E. (1960). Portuguese man-of-war. *Sci. Am.* **202**:158-68.

Mansueti, R. (1963). Symbiotic behavior between small fishes and jellyfishes, with new data on that between the stromateid, *Peprilus alepidotus*, and the scyphomedusa, *Chrysaora quinquecirrha. Copeia.***1**:40-80.

Picchard, J. (1973). *The Sun beneath the Sea.* Hall, London.

Pilsbry, H.A. (1916). The sessile barnacles (Cirripedia) contained in the collections of the U.S. National Museum. *Bull U.S. Nat'l. Mus.* **93.**

Schojima, Y. (1963). Scyllarid phyllosomas' habit of accompanying the jelly fish. *Bull. Jap. Soc. Sci. Fish.* **29**:349-353.

Zahl, P.A. Man-of-war fleet attacks Bimini. In: *The Book of Fishes.* Nat. Geog. Soc. 163-88.

Zann, L.P. (1975). Biology of a barnacle (*Platylepas ophiophilus* Lanchester) symbiotic with sea snakes. In: *The Biology of Sea Snakes.* (Dunson (ed.)). University Park Press, Baltimore.

Zann, L.P. and Harker, B.M. (1978). Egg production of the bar-
    nacles *Platylepas ophiophilus* Lanchester (from seasnakes),
    Platylepas hexastylos (O. Fabricus) (from dugongs), *Octolasmis
    warwicki* Grey (from bay lobsters) and *Lepas anatifera* Lin.
    (from drifting pumice). *Crustaceana.* (in press).

Chapter Seven

Bennett, Isobel (1962). Hydroids on marine snails. *Aust. Nat.
    Hist.* Sept., p. 93.
Bruce, A.J. (1972). Shrimps that live with molluscs. *Sea Frontiers.*
    18: 218-227.
Bruce, A.J. (1972). Notes on some Indo-Pacific Pontoniiae: xx.
    Pontonia sibogae sp. nov., a new species of *Pontonia* from East-
    ern Australia and Indonesia (Decapoda, Natantia, Palaemon-
    idae). *Crustaceana.***23:**182-186.
Cheng, T.C. (1967). Marine molluscs as hosts for symbioses. *Adv.
    mar. Biol.* **5:**1-424.
Christensen, A.M. and McDermott J.J. (1958). Life history and
    biology of the oyster crab *Pinnotheres ostreum* Say. *Biol. Bull.*
    114:146-179.
Dix, T.G. (1973). Mantle changes in the pearl oyster *Pinctada
    maxima* induced by the pea crab *Pinnotheres villosulus. Veliger.*
    15:330-331.
Galtsoff, P.S. (1964). The American oyster *Crassostrea virginica*
    Gmelin. *Fishery Bull. Fish Wildl. Serv. U.S.* **64:**1-480.
Hyman, L. (1967). *The Invertebrates. VI. Mollusca I.* McGraw-
    Hill, New York.
Millar, R.H. (1971). The biology of ascidians *Adv. mar. Biol.* **9:**1-
    100.
Newcomb, E.H., and Pugh, T.D. (in press). Blue-green algae as-
    sociated with ascidians of the Great Barrier Reef. *Science.*
Risbec, J. (1935). Biologie et poute de mollusques gasteropodes
    Neo-Caledoniens. *Bulletin de la Societe Zoologique de France.*
    60:387-417.
Yonge, C.M. (1953). Observations on *Hipponix antiquatus* (Lin-
    naeus). *Proc. California Acad. Sci.* **28:**1-24.

Chapter Eight

Allen, G.R. (1972). Observations on a commensal relationship between *Siphamia fuscolineata* and the crown-of-thorns starfish, *Acanthaster planci. Copeia.* **3**:595-597.

Allen, G.R., and Stark, W.A. II. (1973). Notes on the ecology, zoogeography, and coloration of the gobiesocid clingfishes, *Lepadichthys caritus* Briggs and *Diademichthys lineatus* (Sauvage). *Proc. Lin. Soc. N.S.W.* **98**:95-97.

Bruce, A.J. (1975). Shrimps that live with Echinoderms. *Sea Frontiers.* Jan.-Feb. 1975.

Cannon, L.R.G. (1972). Biological associates of *Acanthaster planci. Report, Crown-of-Thorns Seminar.* Dept. Education and Science, Canberra.

Castro, P. (1971). Nutritional aspects of the symbiosis between *Echinoecus pentagonus* and its host in Hawaii, *Echinothrix calamaris.* In: *Aspects of the Biology of Symbiosis* (Cheng, T.C. (ed.)). University Park Press: Baltimore.

Chesher, R.H. (1970). *Acanthaster planci.* Impact on Pacific coral reefs. *U.S. Dept. Interior Publ.* 187631.

Fishelson, L. (1974). Ecology of the Northern Red Sea crinoids and their epi- and endozoic fauna. *Mar. Biol.* **26**:183-192.

Gibbs, P.E. (1969). Aspects of polychaete ecology with particular reference to commensalism. *Phil. Trans. Roy. Soc.* **B255**:443-458.

Hyman, L. (1955). *The Invertebrates. IV. Echinodermata.* McGraw-Hill, New York.

Randall, J.E. (1964). Notes on the biology of the echinoid *Diadema Antill Arum. Carib. J. Sci.* **4**:421-433.

Smith, C.L. (1964). Some pearlfishes from Guam, with notes on their ecology. *Pac. Sci.* **28**:34-40.

Whitley, G.P. (1950). Clingfishes. *Australian Mus. Mag.* **10**:124-128

Chapter Nine

Crossland, C.J. and Barnes, D.J. (1974). The role of metabolic nitrogen in coral calcification. *Mar. Biol.* **28**:325-332.

Crossland, C.J., and Barnes, D.J. (1976). Acetylene reduction by coral skeletons. *Limnol. Oceanogr.* **21**:153-156.

Hinde, R., and Smith, D.C. (1974). 'Chloroplast symbiosis' and the extent to which it occurs in *Sacoglossa* (Gastropoda: Mollusca). *Biol. J. Linn. Soc. Lond.* **6**:349-356.

Jeffrey, S.W., and Vesk, M. (1976). Further evidence for a membrane bound endosymbiont within the dinoflagellate *Peridinium foliaceum. J. Phycol.* **12**:450-455.

Kremer, B.P. (1975). 14CO²-fixation by the endosymbiotic alga *Platymonas convolutae* within the turbellarian *Convoluta roscoffensis. Mar. Biol.* **31**:219-226.

Lee, J.J., and Bock, W.D. (1976). The importance of feeding in two species of soritid foraminifera with algal symbionts. *Bull. Mar. Sci.* **26**:530-537.

Lewin, R.A. (1976). Prochlorophyta as a proposed new division of algae. *Nature (Lond.)* **261**:697-698.

Muscatine, L. (1974). Endosymbiosis of Cnidarians and Algae. **In:** *Coelenterate Biology.* (Muscatine and Lenhoff (eds.)). Academic Press, New York.

Symbiosis. Symposium of the Society for Experimental Biology. Vol. **29.** Cambridge University Press, Cambridge. (1975).

Taylor, D.L. (1974). Symbiotic marine algae: Taxonomy and biological fitness. **In:** *Symbiosis in the Sea.* (Vernberg (ed.)). Belle W. Baruch Library in Marine Science Vol. 2. Univ. of South Carolina Press, Columbia.

Vandermeulen, J.H., and Muscatine, L. (1974). Influence of symbiotic algae on calcification in reef corals: Critique and progress report. *Ibid.*

Chapter Ten

Allen, G.R. (1972). *The Anemonefishes.* T.F.H. New Jersey.

Blosch (1961). Was ist die Grundlage der Korallenfisch-symbiose: Schutzstoff oder schutzverhalten? *Naturwissenchaften.* **48**:387.

Collingwood, C. (1868). Note on the existence of gigantic sea anemones in the China Sea, containing within them quasi-parasitic fish. *Ann. Mag. Nat. Hist.* **4**:31-33.

Davenport, D. and Norris, K.S. (1958). Observations on the symbiosis of the sea anemone *Stoichactis* and the pomacentrid fish *Amphiprion percula. Biol. Bull.* **115**:397-410.

Duerden, J.E. (1905). On the habits and reactions of crabs bearing actinians in their claws. *J. Zool.* **2**:494-511.

Eibl-Eibesfeldt, I. (1960). Beobachtungen und Versuche an Anemonenfischen (*Amphiprion*) der Malediven und der Nicobaren. *Z. Tierpsychol.* **17**:1-10.

Fishelson, L. (1965). Observations and experiments on the Red Sea anemones and their symbiotic fish *Amphiprion bicinctus. Bull. Sea Fish Res. Stat. Haifa.* **39**:1-14.

Fox, H.M. (1965). Confirmation of old observations on the behaviour of a hermit crab and its commensal sea anemone. *Ann. Mag. Nat. Hist.* **8**:173.

Gohan, H.A.F. (1948). Commensalism between fish and anemone (with a description of the eggs of *Amphiprion bicinctus* Ruppell). *Fouad I. Univ. Publ. Marine Biol. Sta. Ghardaga (Red Sea).* **6**:35-44.

Graefe, G. (1964). Zur Anemonen-Fisch-Symbiose und ihre Grundlage—nach Freilanduntersuchungen bei Eilat/Rotes Meer. *Z. Tierpsychol.* **21**:468-485.

Hadley, D.J. (1973). Ecological studies on the relationships between the anemone *Stoichactis haddoni* (Saville-Kent) and its animal symbionts. M. Sc. thesis (unpubl.) Zoology Dept., University of Queensland.

Mariscal, R.N. (1970). An experimental analysis of the protection of *Amphiprion xanthurus* Cuv. and Val. and some other anemone fishes from sea anemones. *J. exp. mar. Biol. Ecol.* **4**:134-149.

Mariscal, R.N. (1970). The nature of the symbiosis between Indo-Pacific anemone fishes and sea anemones. *Mar. Biol.* **6**:58-65.

Mariscal, R.N. (1971). Experimental studies on the protection of anemone fishes from sea anemones. **In:** *Aspects of the Biology of symbiosis.* (Cheng (ed.)). University Park Press, Baltimore.

Robertson, R. (1963). Wentletraps (Epitoniidae) feeding on sea anemones and coral. *Proc. Mal. Soc. Lond.* **35**:51-63.

Ross, D.M. (1974). Evolutionary aspects of associations between crabs and sea anemones. **In:** *Symbiosis in the sea.* (Vernberg (ed.)). University of South Carolina Press, Columbia.

Chapter Eleven

Doak, W. (1975). *Sharks and Other Ancestors.* Hodder and Stoughton, Auckland.

Munro, I.S.R. (1967). *The Fishes of New Guinea.* Dept. Agriculture, Stock and Fisheries, Port Moresby.

Strasburg, D.W. (1959). Notes on the diet and correlating structures of some central Pacific echeneid fishes. *Copeia.* **3**:244-248.

Strasburg, D.W. (1964). Further notes on the identification and biology of echeneid fishes. *Pac. Sci.* **18**:51-57.

Whitley, G.P. (1949). Sucking fishes. *Aust. Mus. Mag.* **13**:17-23.

Whitley, G.P. (1951). The satellite of sharks. *Aust. Mus. Mag.* **15**:1-154.

Chapter Twelve

Beebe, W. (1924). *Galapagos: World's End.* Putnam, New York.

Beebe, W. (1928). *Beneath Tropic Seas.* Putnam, New York.

Bruce, A.J. (1975). Coral reef shrimps and their colour patterns. *Endeavour.* **34**:23-27.

Eibl-Eibesfeldt, I. (1961). Eine Symbiose zwischen Fischen (*Siphamia versicolor*) und Secigeln. *Z. Tierpsychol.* **18**:56-59.

Feder, H.M. (1966). Cleaning symbiosis in the marine environment. **In:** *Symbiosis* (Henry (ed.)). Academic Press: New York.

Fishelson, L. (1974). Ecology of the northern Red Sea crinoids and their epi- and endozoic fauna. *Mar. Biol.* **26**:183-194.

Hobson, E.S. (1969). Comments on generalizations about cleaning symbiosis in fishes. *Pac. Sci.* **23**:35-39.

Limbaugh, C. (1961). Cleaning symbiosis. *Sci. Am.* **205**:42-49.

Limbaugh, C., H. Pederson, and F.A. Chace. (1961). Shrimps that clean fishes. *Bull. Mar. Sci.* **11**:237-257.

Losey, G.S. (1971). Communication between fishes in cleaning symbiosis. **In:** *Aspects of The Biology of Symbiosis* (Cheng (ed.)). University Park Press, Baltimore.

Losey, G.S. (1974). Cleaning symbiosis in Puerto Rico with comparison to the tropical Pacific. *Copeia.* **4**:960-970.

Randall, J.E. (1958). A review of the labrid fish genus *Labroides*, with descriptions of two new species and notes on ecology. *Pac. Sci.* **13**:327-347.

Randall, J.E. (1962). Fish service stations. *Sea Frontiers.* **8**:40-47.

Robertson, R. (1973). Sex changes under the waves. *New Scientist.* May, pp. 538-540.

Sargent, R.C. and Wagenbach, G.E. (1975). Cleaning behaviour of the shrimp *Periclimenes anthophilus* Holthuis and Eibl-Eibesfeldt (Crustacea:Decapoda:Natantia). *Bull. mar. Sci.* **25**: 466-472.

Youngbluth, M.J. (1968). Aspects of the ecology and ethology of the cleaning fish *Labroides phthirophagus* Randall. *Z. Tierpsychol.* **25**:915-932.

Wickler, W. (1968). *Mimicry in Plants and Animals.* World University Library, London.

INDEX

Page numbers given in *italics* refer to illustration.

411

416